Light Power:
Half a Century of Solar Electricity Research

Volume 2: 20th Century Photovoltaic Systems

Light Power:
Half a Century of Solar Electricity Research

Volume 2: 20th Century Photovoltaic Systems

editor

David Faiman

Ben-Gurion University of the Negev, Israel

World Scientific

NEW JERSEY · LONDON · SINGAPORE · BEIJING · SHANGHAI · HONG KONG · TAIPEI · CHENNAI · TOKYO

Published by

World Scientific Publishing Co. Pte. Ltd.

5 Toh Tuck Link, Singapore 596224

USA office: 27 Warren Street, Suite 401-402, Hackensack, NJ 07601

UK office: 57 Shelton Street, Covent Garden, London WC2H 9HE

British Library Cataloguing-in-Publication Data
A catalogue record for this book is available from the British Library.

LIGHT POWER: HALF A CENTURY OF SOLAR ELECTRICITY RESEARCH
Volume 2: 20th Century Photovoltaic Systems

ISBN 978-981-123-131-5 (hardcover)
ISBN 978-981-123-132-2 (ebook for institutions)
ISBN 978-981-123-133-9 (ebook for individuals)

For any available supplementary material, please visit
https://www.worldscientific.com/worldscibooks/10.1142/12134#t=suppl

Typeset by Stallion Press
Email: enquiries@stallionpress.com

https://doi.org/10.1142/9789811231322_fmatter

Preface to Volume 2

History, as pointed out in Volume 1, is usually written by people who did not themselves witness the events they document. On those occasions in recent times when it has been possible to witness events of historic value, it is often non-specialists, such as journalists, who perform the documentation. This latter fact has its virtues in that a more balanced picture may result than might otherwise have been the case had the history been written by a single participant. However, the downside of a non-specialist history is frequently a lack of important details.

The advent of large-scale solar power generation is certainly a historic event that has come about within living memory and one which continues to develop as these words are being written. Furthermore, the editor, although not himself having contributed anything of historic value to the subject, has been intimately involved in the research and development of solar power, and has been fortunate to have associated with many of the scientists and industrialists who have brought it about. In particular, for a quarter of a century, he hosted a sesquiennial symposium on solar power generation to which were invited many of the most influential people whose creations indeed led to the present situation.

Balance was provided in those symposia by the inclusion of all solar power technologies within a single meeting, lasting two or three days, with no parallel sessions. That is to say, a unique feature of the *SedeBoqer Symposia* (referring to the venue at which they were held) was that all participants, no matter what their particular specialty was, were able to attend all presentations. This fact placed an important constraint on the keynote speakers: Namely, that the first 15 minutes of their otherwise specialist presentation had to be in a language that all could understand. "All" in this situation referred to scientists from other disciplines, but equally important, to industrialists and government decision makers. Those keynote lectures, which were originally published in booklets by Ben-Gurion University of the Negev, and distributed, some months later, to all the attendees, thus constitute a record of progress, in the actual words of the people who were responsible for much of it, and, equally important, in a format that is accessible to a wide range of readers.

This was the motivation that led the present editor to republish a selection of the keynote lectures in the form of a history via the words of those who made it. The original idea was to reproduce those lectures in strict chronological order, maintaining the year-by-year mix of photovoltaic and solar-thermal technologies, in order to emphasize the manner in which interest in each has ebbed and flowed throughout the past half century. However, it turned out that doing so would have rendered such a book unreasonably lengthy. So, instead, a separation was made between the solar-thermal presentations, which constitute Volume 1 of this history, and the photovoltaic (PV) presentations. This volume (Volume 2) covers PV developments through to the end of the 20th century. A forthcoming volume (Volume 3) will take us into the 21st century. It makes sense to create a separation between large-scale solar thermal and PV power production because the former got off to an earlier start, and is now relatively mature. On the other hand, terrestrial PV, which started with flat panels of extremely costly silicon cells with low efficiency and questionable reliability, has evolved into more affordable panels of useful efficiency and high reliability. Moreover, the PV field has further branched out into what appears to be a never-ending exploration of new system ideas and materials: an exploration that continues to the present day and which shows no signs of diminishing.

Thus, to summarize: The present volume presents a specialist history of 20th century PV power generation, edited by a specialist, but with balance maintained by including the opinions of many of the competing specialists (including the solar-thermal variety) whose creations led to the present commercialization of the field. Those opinions are reflected both in the presentations themselves and in the recorded discussion sessions that are reproduced at the end of most presentations. As already emphasized, interested non-specialists should be able to follow this history by virtue of the special constraint that was placed upon the speakers whose keynote lectures constitute the bulk of all three volumes.

Acknowledgments

The editor acknowledges Ben-Gurion University of the Negev, the copyright-holders of the symposium proceedings volumes, from which most of the material in this book has been extracted. He is deeply indebted to Ms. Roxana Dann, for scanning the pre-digital presentations from the older among those volumes, converting the scans to editable files, and performing a first proof-reading of the entire contents of this book. He thanks Elsevier for use of some material in chapters 4 and 7. He thanks: Keith Barnham, Keith Emery, Vahan Garboushian, Michael Grätzel, Antonio Luque, Morton Prince, Vern Risser, and Serdar Sariciftci for their agreement to have their historic presentations reproduced in this volume. The historic presentations by Stephen Kaneff, Mark Koltun, Peter Landsberg, and Fred Treble are reproduced here without permission as their authors are sadly no longer with us. May these contributions act as tributes to their memory. Photo credits are due to: Gerald and Buff Corsi © California Academy of Sciences, for the cover picture; Franco Roco and Salvatore Castello of ENEA, for Fig. 2.1, Jeanine and Noel Cotter of Luminalt, for Fig. 2.3. More generally, the editor wishes to thank the staff of World Scientific Publishers for the excellent work they have performed on restoring several of the old figures that are reproduced in this book. Finally, every effort has been made to identify and acknowledge the copyright holders of the many figures that are reproduced within the presentations of the various contributors to this book. Where errors have occurred, the editor offers his profound apologies and will include any requested errata in Volume 3.

Contents

Chapter 8. 1999   **275**

Afterword   313

Introduction

Life in the developed countries has come to depend upon electric power being available all the time. The principal means for generating this power was inherited from an earlier industrial age when power was provided by wood, coal, gas and oil. The fact that those fuels produced heat energy led to the development of heat engines, machinery for industrial manufacture and transportation, and the science of thermodynamics. With the discovery of electricity, it was natural to employ heat engines to drive the turbines for its generation. The great advantage of electric power compared to the direct use of, say coal or oil, was that it could be generated at a single location and transported great distances almost instantaneously: by metal cables to distant factories, shops, schools, homes, etc., and by the tracks themselves for trains. Thus, instead of needing to carry coal between stations where its fuel load would need to be replenished, an electric locomotive could travel non-stop, its motor being fed *en route* by precisely the amount of electricity it needed. Electric trains were also cleaner than the coal-powered variety, making travel more pleasant for passengers and life cleaner for people living close to the tracks.

With the successively higher levels of efficiency that the science of thermodynamics enabled heat engines to reach, it was a natural step to seek alternative fuels that could provide heat to generate electricity more cleanly than could fossil fuels. This challenge led to the development of nuclear heat and then solar heat. It may, at first, seem surprising that a technology as complex as a thermo-nuclear reactor would have achieved industrial-scale development prior to that of a far simpler source of heat, namely the sun. But it must be remembered that the scientific and engineering knowledge for nuclear power had been developed during World War II when both of the opposing sides were in a race towards achieving nuclear weapons before the other. So, like many other spin-offs from war technologies, such as plastics and radar, nuclear power became ripe for peace-time exploitation.

This is not an appropriate forum for discussing the pros and cons of nuclear power, as our point is to emphasize that the development of solar-thermal power was in a sense a natural evolution from fossil fuels for the generation of electricity. Specifically, the heat generated by a field of solar collectors could be fed directly into an existing infrastructure.

By contrast, the development of photovoltaic technology, which converts sunlight directly to electricity *without a thermal intermediate stage*, represents a less comfortable stage in the evolution of power generation. Volume 1 in this three-volume history covered the principal details of solar-thermal development. The present volume will present the principal 20th century developments in the evolution of photovoltaic power plants.

There is of course one major problem that must not be forgotten whether one wants to use solar energy as a heat source for turbo-generation or photovoltaic cells for direct conversion, namely, that solar power on Earth is an enormously dilute source compared to fossil fuel.

As an indication of how dilute it is, a receiver the size of a barrel of oil, collecting solar energy in a sunny desert region at 100% efficiency, would take approximately one year to collect the same amount of energy that is contained in a barrel of oil! This means that solar-powered replacements of fossil-fed power stations have to occupy large areas of ground if located on the Earth. How large? If a 100 MW oil-powered electricity-generating station has a capacity factor CF, then in one year it will generate

$$100\,\text{MW} \times 24\,\text{hr/day} \times 365\,\text{days/year} \times CF = 876,000,000 \times CF \text{ kWh}$$

of electrical energy. Capacity factors of oil plants are usually between 10 and 20%. So taking $CF = 0.15$ in our example, we expect such an oil-fired plant to generate 131,400,000 kilowatt hours of electricity annually. How many barrels of oil does this require? A typical barrel of oil contains about 1,700 kWh of useable thermal energy, which can be converted to electricity at about 30% efficiency. Thus one barrel of oil can provide for about 510 kWh of electricity. Hence our 100 MW power plant would consume about 257,647 barrels in the course of a year. Finally, if the effective cross-sectional area of a barrel-size solar receiver is about $1 \, m^2$, then, at the impossibly high value of 100% conversion efficiency, our solar collector would have to be $257,647 \, m^2$ in area, i.e. about one-quarter of a square kilometer. Since a realistic annual solar to electrical efficiency is more like 10%, then a solar plant that generates the same annual output as a 100 MW oil-burning plant would need to occupy approximately $2.5 \, km^2$ of land.

During the year 2018, the world generated 26,614.8 TWh of electricity [1]. From the above order-of-magnitude estimate, it would require about $500,000 \, km^2$ of land to generate that much energy from sunlight. This *is* about one-twentieth the area of the Sahara Desert. Of course, it need not all be located in the Sahara. That desert could provide all of the electrical needs of the African continent and Europe, while the other deserts in the world could provide for other regions [2–6].

Thus, contrary to the claims of many nay-sayers, there *is* enough collection area in the deserts of the world to provide for all of the solar-generated electricity that humankind may need for a long time to come. And thereafter? One need not resort solely to the creativity of science fiction writers to realize that research is already underway into methods for future power to be generated by Earth-orbiting satellites, or even on the Moon followed by transmission to the Earth via a variety of cableless possibilities.

Of course, the large-scale *generation* of solar power is not the only challenge that must be overcome. A network or networks of low-loss transmission methods must be installed worldwide and appropriate storage systems must be developed to enable the solar-generated power to be managed efficiently at points of consumption. So the transition from fossil to solar power cannot be achieved overnight. On the other hand, much of the world has a well-developed fossil-powered infrastructure that can act as a backup for a gradually expanding solar input — an expansion that should increase as existing fossil-powered plants reach their end-of-life economic projections and are replaced by solar or wind plants.

But returning to the present and how we have arrived here, the history of solar photovoltaic power, as represented in this series by the Sede Boqer Symposia, is conveniently divided into two parts. The present volume (the second in the entire series) covers developments through the year 1999. The third volume will cover some of the significant developments in the early part of the 21st century.

Twentieth century terrestrial solar plants were powered by silicon cells, European systems being reviewed here by Fred Treble, and US systems by Vern Risser.

Largely because of the then still high cost of silicon cells, many ingenious methods were developed for minimizing the use of that material via the concentration of sunlight. The presentations of Vahan Garboushian, Steve Kaneff and Antonio Luque are typical of some of the successes back then.

Nevertheless, there was already much research into alternative materials as witnessed by the presentations of Keith Barnham, Keith Emery, Michael Grätzel, Mark Koltun, Morton Prince, and Serdar Sariciftci.

However, whatever material or materials the future will decide upon, the fundamental constraints that the laws of physics place upon conversion efficiency will have to be obeyed. This subject is extensively reviewed in the presentation of Peter Landsberg.

References

[1] *BP Statistical Review*, 2019.
[2] K. Kurokawa (ed.), *Energy from the Desert: Feasibility of Very Large Scale Photovoltaic Power Generation (VLS-PV) Systems*, James & James, 2003.
[3] K. Kurokawa, K. Komoto, P. van der Vleuten, and D. Faiman (eds.), *Energy from the Desert: Practical Proposals for Very Large Scale Photovoltaic Systems*, Earthscan, 2007.
[4] K. Komoto, C. Breyer, E. Cunow, K. Megherbi, D. Faiman, and P. van der Vleuten (eds.), *Energy from the Desert: Very Large Scale Photovoltaic Systems; Socio-economic, Financial, Technical and Environmental Aspects*, x Earthscan, 2009.

[5] K. Komoto, C. Breyer, E. Cunow, K. Megherbi, D. Faiman, and P. van der Vleuten (eds.), *Energy from the Desert: Very Large Scale Photovoltaic Power – State of the Art and Into the Future.* Routledge, 2013.

[6] *Energy from the Desert: Very Large Scale PV Power Plants for Shifting to a Renewable Energy Future.* International Energy Agency Photovoltaic Power Systems Program (IEA-PVPS Task 8, External Final Report, February 2015, ISBN 978-3-906042-29-9).

Chapter 1

Photovoltaic Power from the Perspective of 1986

1.1 Editor's Foreword

Volume 1 of this history started with a review of the entire solar power situation as it stood at the time of the *1st Sede Boqer Symposium on Solar Electricity Production*, in February 1986. There were no invited keynote presentations, because that series of symposia had not yet developed a canonical structure. Instead, the present editor, as host of the symposium, invited himself to present a review of the then state-of-the-art. That review, even though it included several photovoltaic systems, was reproduced in its entirety in Volume 1, which was devoted to solar thermal plants. It is worth recalling that at that time, the largest solar-thermal plant, which had recently started operation, was SEGS-2 with a rating of 30 MWe, in Daggett, California. By comparison, large PV plants had ratings two orders of magnitude less, by far the largest being the 5 MWp Plant (depicted on the cover of this volume) at Carrisa Plain, California.

In principle that same 1986 review could provide a useful starting point for the present volume. However, not wanting to be unnecessarily repetitious, while at the same realizing that not all present readers will be familiar with the solar-thermal volume, the PV sections have been extracted from that review and will constitute the remainder of the present chapter. One further omission both here and in Volume 1 is the absence of the discussion that followed this presentation, because no recording was made — a practice that became standard in later symposia.

Finally, the reader should not be confused by the fact that the figures and references have been re-numbered in the present chapter compared to their original numbering in Volume 1. Other than this change, the text is identical to its format in the previous volume.

1.2   There Was Solar Power before 1986 (Prof. David Faiman)

Photovoltaic extracts from the full keynote review presented by David Faiman (Ben-Gurion University, Sede Boqer Campus, Israel), reproduced in volume 1.

1.2.1   *Photovoltaic systems*

The great attraction of photovoltaic cells as a source of solar generated electricity resides in their great simplicity: A photovoltaic station need not contain any moving parts. There is also no inertia in a photovoltaic system since the cells respond instantly to illumination. This point is well illustrated in Figs. 1.1 and 1.2 which

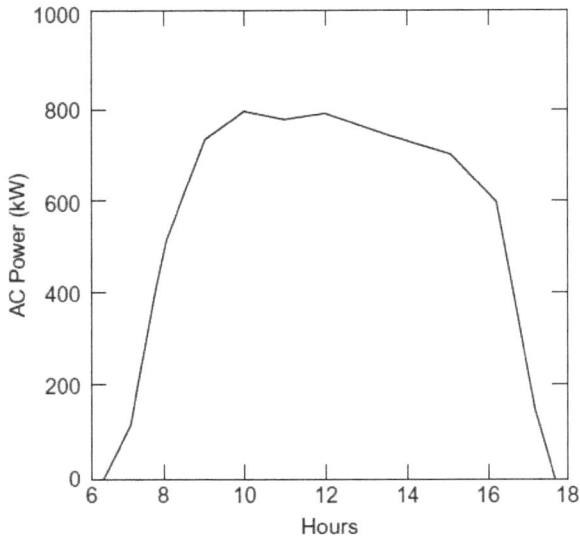

Figure 1.1:   Output power of the Lugo photovoltaic system throughout a typical day.

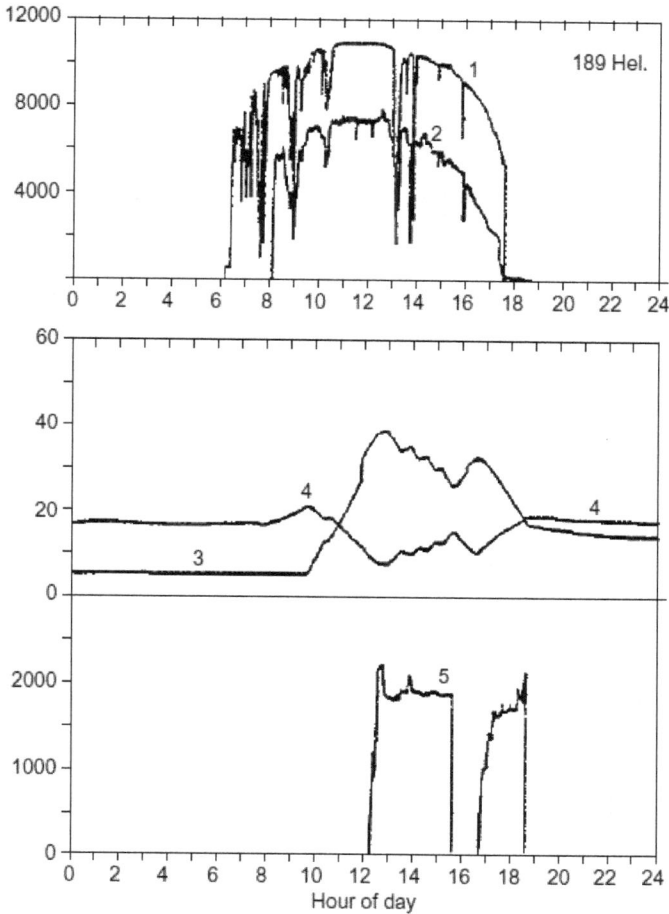

Figure 1.2: Output power of the Themis solar-thermal system throughout a typical day.

show, respectively, the electric power output of the Lugo photovoltaic system and the Themis solar thermal facility.

In both examples sunrise is seen to occur at about 6:30, but whereas the Lugo system has reached maximum power output by about 10:00 am (having started to supply electricity at sunrise), Themis does not begin to feed into the grid until about 12:30 pm.

Another attractive feature of photovoltaic power stations is the speed with which they can be installed. In fact, at the time this review was presented orally, I was able to list the existing photovoltaic facilities in a table similar to Tables 1.2.3, 1.2.6 and 1.2.7. Now (at the time of writing up the review), their number exceeds 20. Therefore, inclusion in a table is no longer feasible. Instead, I shall list them singly, drawing attention to the major features. Again, as with the solar thermal facilities, my emphasis will be on the negative aspects. It must be kept in mind however, that, in most cases, the systems have not been operating long enough for problems to have been fully documented.

1.2.1.1   *Bridges national monument (UT), USA*

Installed in 1979, this was one of the first "large" photovoltaic power projects. The system comprises 1,700 sq.m of photovoltaic cells manufactured by three companies: Motorola, Arco Solar and Spectrolab. It supplies most of the electricity for a remote National Park which was previously served entirely by a diesel-powered generator. The system is reported to have cost $31/Wp.

References: [1, 2].

1.2.1.2   *Kythnos Island, Greece*

System comprises 800 Siemens type SM 144-9 modules. Each module consists of 144, 10 cm diameter, monocrystalline silicon cells. The array has an area of 1,200 sq.m and a nominal rating of 100 kWp. The system entered operation in 1982.

References: [3, 4].

1.2.1.3   *Hamamatzu City, (Shizuoka Pref.) Japan*

A 100 kWp facility that provides the initial charging of batteries produced by an automobile battery factory.

References: [5, 6].

1.2.1.4   *San Augustin de Guadalix, nr. Madrid, Spain*

A 100 kWp central power station based, largely, on bi-facial singe-crystal cells manufactured by the Isofoton corporation. All modules are stationary, flat-plate types and the entire facility occupies 7,000 sq.m.

Reference: [7].

1.2.1.5 *Puerto Rico*

A 100 kWp stationary flat-plate facility.

1.2.1.6 *Caltrans, California, USA*

A 100 kWp stand-alone system which complements three existing diesel generators of the California Department of Transport. Modules are Arco Solar "square" single-crystal cells.

Reference: [8].

1.2.1.7 *Shopping center, Lovington (NM), USA*

A 104 kWp flat-plate array. The system is reported to have cost $30/Wp.

Reference: [2].

1.2.1.8 *Oklahoma center, Oklahoma City (OK), USA*

System consists of 1,512 Solarex 65 cm × 128.5 cm modules with a nominal rating of 106 kWp but augmented to 140 kWp by stationary plane mirrors. Each module contains 72 polycrystalline 10 cm × 10 cm cells.

System went into operation in 1982, surplus electricity is sold to the utility. Total system cost was $2.98 M (about $21/Wp). Output power is lower than expected: low module output is suspected of being the cause.

References: [2, 9].

1.2.1.9 *John Long 24-house project, Phoenix (AZ), USA*

A housing project supplied by a 192 kWp stationary array of Arco Solar "square" single-crystal photovoltaic cells. The system is grid — connected to the local utility, which purchases surplus power in the following way.

The utility's buying-selling prices are: 0.0503–0.0743 \$/kWh during the period 12:00 to 22:00 and 0.0265–0.0293 \$/kWh during off-peak hours.

The system is reported to have cost about \$10/Wp.

References: [8, 10].

1.2.1.10   *Tsukuba University, (Ibaraki Pref.), Japan*

A 200 kWp array supplying electricity for lighting and laboratory equipment.

References: [5, 6].

1.2.1.11   *Ichihara City, (Chiba Pref.), Japan*

A grid-connected, distributed system comprising a total of 200 kWp of cells on rooftops and other available space.

References: [5, 6].

1.2.1.12   *Solarex factory, Frederick (MD), USA*

2,500 sq.m of roof-mounted Solarex polycrystalline cells (nominal rating: 200 kWp) supply all the electricity requirements for this solar cell manufacturing "breeder" plant.

1.2.1.13   *Sky Harbor airport, nr. Phoenix (AZ), USA*

This system consists of 80 two-axis trackers. Each tracker comprises 272 individual solar cells each placed at the focus of an R = 33 × fresnel lens. The fresnel lenses have apertures of approximately 30 cm × 30 cm and the power rating of the entire system is 225 kWp. The trackers were built by Martin Marietta Corp. The system has been operational since 1982 and is reported to have cost \$11/Wp.

References: [2, 11, 12].

1.2.1.14   *Pellworm Island, West Germany*

The system consists of a (mainly) stationary flat-plate array of 17,568 AEG-Telefunken type PQ 10/20/0 modules. Each module contains

20 multi-crystalline cells of dimensions 10 cm × 10 cm. The nominal power rating is 300 kWp. The system was installed in 1983 at a reported cost of about \$15.6/Wp. It occupies a total land area of 16,500 sq.m.

References: [4, 13, 14].

1.2.1.15   *Georgetown Univ. Intercultural center, Washington DC, USA*

This system consists of stationary, architecturally integrated, flat plate modules of nominal power 300 kWp. The modules contain Solarex microcrystalline cells and occupy about 3,300 sq.m. The system became operational in 1984. It is reported to have cost \$21/Wp.

Reference: [2].

1.2.1.16   *Mississippi County Community College, Blythesville (AK), USA*

System consists of 270, N-S axis, parabolic trough collectors made by Solar Kinetics Corp, each having an aperture area of 13 sq.m and a concentration ratio of about R = 42×. The collectors concentrate light onto linear modules of single-crystal photovoltaic cells made by Solarex. The cell modules are mounted on a liquid-cooled heat sink which enables the system to supply about 22 GJ per day (at peak output) of thermal energy, in addition to 320 kWp of electrical power. The photovoltaic cells are 2.5 cm × 5.0 cm in size, and operate with an efficiency of about 12% at 55°C.

The system became operational in 1981 but the electricity-generating part could not be made to function in a trouble-free manner. It is reported to have cost \$30/Wp. The last published DOE summary on the project (dated October 1983) mentioned investigations that were underway to convert the system from a photovoltaic to a thermal facility.

References: [2, 15, 16].

1.2.1.17   *City of Austin project, (TX), USA*

A 300 kWp project of the local utility, for city power generation.
Reference: [8].

1.2.1.18   *The Soleras project, Saudi Arabia*

This system is similar to that at Sky Harbor airport in the USA, but larger. Its rated power output is 350 kWp. There are 160 arrays, each containing 32 cell modules. The project, which is well-documented, was installed in 1981 and the only serious problem appears to have been the gradual failure (as of August 1984) of 6–7% of the modules. Module degradation has been attributed to manufacturing faults rather than environmental effects. The system is reported to have cost \$53/Wp.

References: [2, 17–19].

1.2.1.19   *"Lugo" power station, nr. Hesperia (CA), USA*

This system comprises 108 two-axis trackers each containing 256 Arco Solar flat-plate modules of "round" single-crystal photovoltaic cells. The project occupies a land area of 20 acres and has a nominal power rating of 1 MW. Reference 20 provides a valuable indication of the various sources of power loss associated with this plant. According to Arnault *et al.* the conversion of the dc output to ac involves a total loss of 18%, which breaks down as follows:

Module/panel interconnects:	0.7%
Field wiring:	1.3%
Inverter and switching gear:	4.0%
Power for tracking:	3.0%
Shadowing:	2.5%
Module mismatch:	3.0%
Nighttime ac standby:	2.0%
Operational downtime:	1.5%

References: [20–23].

1.2.1.20 *Centralized power station, Saijo (Ehime Pref.), Japan*

This 1 MWp system consists of stationary, flat-plate modules made by various photovoltaic cell manufacturers. It started to produce 200 kWp in 1982 and was enlarged step-by-step to 1,000 kWp in 1986. According to Ref. [6], "It is difficult to find a wide space for such a centralized plant in Japan."

References: [5, 6].

1.2.1.21 *S.M.U.D., Rancho Seco nr. Sacramento (CA), USA*

This central power facility, operated by the Sacramento Municipal Utility District, comprises two subsystems each of nominal power 1.2 MWp. The arrays are Arco Solar single-crystal modules mounted on Acurex single (N–S) axis trackers. The two subsystems constitute the first two phases of what was planned to be a 100 MWp facility, but for the time being (DOE summary, October 1985) all subsequent phases have been cancelled. The present system is reported to have cost $10/Wp.

References: [2, 8].

1.2.1.22 *Carrisa Plain facility, nr. Los Angeles (CA), USA*

Currently the largest photovoltaic facility, with a power rating of 6.5 MWp, this system is similar in principle to Lugo, 90% of the trackers being of the same (33 ft × 36 ft) aperture area but enhanced by plane side mirrors set at an angle of 120° to the plane of the cell modules. The remaining 10% of the trackers are not mirror-enhanced but merely larger (40 ft × 40 ft). Carrisa Plain employs Arco Solar's "square" single-crystal cells.

References: [21, 24].

1.2.2 *References*

[1] E.F. Lyon, "A 100 kW peak photovoltaic power system for the Natural Bridges national monument", in K.W. Boer and B.H. Glen (eds.), *Sun II*, Pergamon, New York, 1979, p. 1732.

[2] L.M. Magid, "U.S. photovoltaic systems experience: Prospects for the future", in *Proceedings of the 17th IEEE Photovoltaic Specialists Conference*, Kissimmee, FL, May 1984.

[3] H. Ritter, "Kythnos photovoltaic power plant", in *Solar Energy R&D in the European Community, Series C, Volume 1, Photovoltaic Power Generation: Proc. of Final Design Review Meeting on EC Photovoltaic Pilot Projects, held in Brussels, 30 Nov–2 Dec*, 1981 ed. W. Palz (R. Reidel Publishing Company, Dordrecht, Holland), p. 217.

[4] F.C. Treble, "The CEC photovoltaic pilot projects", in *Proceedings of the 6th European Community Photovoltaic Solar Energy Conference*, London, April 1985.

[5] K. Kurokawa, H. Akabane, H. Hosokawa and K. Murakami, "A series of R&D on solar photovoltaic conversion systems of Japan", in *Solar World Forum*, eds. D.O. Hall and J. Morton (Pergamon, Oxford, 1982), p. 2755.

[6] T. Horigome, "Solar energy development in Japan" (15 pp typewritten report, dated March 1986), private communication from the author.

[7] J.S. Solera *et al.*, "100 kW experimental photovoltaic power plant", in *Proceedings of the 6th E.C. Photovoltaic Solar Energy Conference*, London, April 1985.

[8] G.J. Shushnar and R. Yenamandra, "Trends in recent applications of large photovoltaic systems", in *Proceedings of the ASME Solar Energy Conference*, Anaheim, CA, April 1986, p. 184.

[9] Y.P. Gupta, "Oklahoma Center for Science and Arts: Intermediate photovoltaic system application experiment, Phase II — Final Report" (DOE/ET/20630-T1), January 1984.

[10] S. Andrews, "A solar developer breaks new ground", *Solar Age* 11 (December 1985/January 1986), p. 40.

[11] L.O. Herwig, "Technology and economic status of concentrating-photovoltaic systems in the United States", in S.V. Szokolay (ed.), *Solar World Congress*, Pergamon, Oxford 1984, p. 1565.

[12] W.J. McGuirk, "Fabrication and construction of the Sky Harbor Airport solar photovoltaic concentrator project, Phase II — Progress Report from 11 March 1980–30 June 1982" (DOE/ET/20624-1.60), January 1984.

[13] H.J. Lowalt, "300 kW photovoltaic pilot plant Pellworm", in *Solar Energy R&D in the European Community, ... (Op. cit.)*, p. 179.

[14] H.J. Lowalt and B. Proetel, "The 300 kW Pellworm solar power station: Performance and experience", in *Proceedings of the 6th E.C. Photovoltaic Solar Energy Conference (Op. cit.)*.

[15] E.M. Henry and H.V. Smith, "Mississippi County Community College solar photovoltaic total energy project", in *Sun II (Op. cit.)*, p. 1729.

[16] H.V. Smith, "The Mississippi County Community College large-scale-demonstration project: A success story", in *Proceedings of the 4th European Community Photovoltaic Energy Conference*, 10–14 May 1982, Stresa, Italy, p. 94.

[17] F. Huraib, B. Khoshaim, A. Al-Sana, M.S. Imamura and A.A. Salim, "Design, installation and initial performance of 350-kW photovoltaic power system for Saudi Arabian villages", in *Proceedings of 4th E-C Photovoltaic Solar Energy Conference (Op. cit.)*, p. 57.

[18] F. Huraib, B. Khoshaim, A. Al-Sana, M.S. Imamura and A.A. Salim, "SOLERAS photovoltaic power systems project: Final report; photovoltaic power seminar February 20–23 1983" (Midwest Research Institute, Kansas City, report No. MRI/SOL 0102).

[19] F.S. Huraib, M.S. Imamura, A.A. Salim and N. Rao, "SOLERAS photovoltaic power systems project: Interim report; module failure analysis" (Midwest Research Institute, Kansas City, report No. MRI/SOL 0101), October 1984.

[20] R.J. Arnault, E. Berman, C.F. Gay, R.E.L. Tolbert and J.W. Yerkes, "The ASI one-megawatt photovoltaic power plant", in *Solar World Congress (Op. cit.)*, p. 1624.

[21] J.C. Arnett and R.F. Reinoehl, "Installation and performance of Lugo and Carrisa Plain photovoltaic systems", in *Proceedings of the 166th Electrochemical Society Meeting*, October 1984.

[22] N.W. Patapoff and D.R. Mattijetz, "Utility experience with an operating central station MW-sized photovoltaic plant", in *IEEE Transactions on Power Apparatus and Systems*, August 1985, p. 2020.

[23] N.W. Patapoff Jr, "Two years of inter-connection experience with the 1 MW at Lugo", in *Proceedings of the 18th IEEE Photovoltaics Specialists Conference*, Las Vegas, NV, October 1985.

[24] T. Hoff and G. Shushnar, "Two years of performance data for the world's largest photovoltaic power plant", in *Proceedings of the IEEE Power Engineering Society Meeting*, Summer 1986.

Chapter 2

1988

2.1 Editor's Foreword

March 1988 saw what was actually the *3rd Sede Boqer Symposium on Solar Electricity Production*, the previous symposium in this series having been heavily dominated by interest in the multimegawatt solar-thermal power plants that had recently begun to produce performance data. (For details, see volume 1 of this history.) Attendees at the third symposium were privileged to receive a keynote address from the late Fred Treble (1916–2010), who had played a major role in the development of test methods for quantifying the performance of photovoltaic systems. Mr. Treble presented details of the 16 pilot PV plants that had participated in the first European study of this technology. (Note: The editor was unable to find a photograph of the *Aghia Roumeli* plant that was illustrated in the original symposium proceedings. Instead, Fig. 2.1 shows the *Vulcano* plant in a photograph that was kindly provided by Franco Roco and Salvatore Castello of ENEA.)

2.2 Experience with photovoltaic pilot plants in Europe

A keynote lecture presented by Mr. F.C. Treble, Consultant (43, Pierrefondes Avenue, Farnborough, Hants., GU14 8PA, United Kingdom).

2.2.1 *Abstract*

This paper presents an updated overview of the 16 photovoltaic pilot plants sponsored by the Commission of the European Communities (DG XII), with special reference to the problems encountered and lessons learned. Details are given of the Commission's plans for tackling these problems and improving the plants in the light of experience.

2.2.2 *Introduction*

As part of their second four-year solar energy R&D program (1979–1983), the Commission of the European Communities (DG XII) sponsored 16 photovoltaic pilot plants in the intermediate range 30–300 kWp. None of the plants is yet in the "commercially viable" category but the basic aim was to show that the direct generation of electricity from the sun is feasible in Europe and can serve a wide range of useful purposes. Specific objectives were as follows:

- To develop systems and critical components for a variety of applications and environments.
- Compare the different technological approaches in terms of their performance, reliability and durability.
- Provide data for improving the system design.
- Determine operational and maintenance requirements.
- Resolve any problems of integration with the grid.

The plants were constructed by consortia of private companies, electricity authorities, universities, government organizations and regional agencies from all over the European Community. The total cost was about 22 million European Units of Account (roughly US$22 million), of which about one third was borne by the Commission and the rest was contributed by industry, member governments and other institutions.

The program involved nine European manufacturers in the production of over 1 MWp photovoltaic modules in sizes ranging from 19 to 120 Wp. For design qualification, prototypes and production samples of each type were subjected to rigorous performance and environmental tests [1] at the CEC Joint Research Centre (JRC), Ispra, Italy.

No design qualification tests were carried out on the other system components, but power conditioning units, such as inverters and converters, were required to undergo acceptance tests to establish their efficiency/load characteristics and check their safety and protective features.

The Preliminary, Intermediate and Final Design Reviews in 1981, with the periodic Contractors' Meetings which followed, provided valuable opportunities for the participants to meet, review the progress and discuss common problems.

Much has already been written on the pilot plants [2–7]. The purpose of the present paper is to update this information in the form of an overview, review the problems encountered and lessons learned and give some details of the program of improvements recently initiated by the Commission.

2.2.3 *The pilot plants*

Brief descriptions of the 16 plants, starting with the five rural electrification projects, are given in the following sections. The consortium leader and type(s) of modules used are listed in parenthesis after each heading.

2.2.3.1 *Aghia Roumeli (SERI-Renault with France-Photon 72 Wp modules)*

A 50 kWp PV generator with a 480 kWh battery supplies AC power to homes, shops, hotels and street lighting in a village of 105 inhabitants on the south coast of Crete. The village can be reached only by boat or by a 5-hour walk through a rocky canyon. A 50 kVA diesel generator provides backup but cannot be operated in parallel with the PV system.

2.2.3.2 *Kaw (SERI-Renault with France-Photon 72 Wp modules)*

A 35 kWp PV system with a 480 kWh battery supplies AC power to a remote village of 70 inhabitants in French Guyana. The electricity is used for homes, a school, a church and street lighting. A diesel generator provides backup, as in Aghia Roumeli.

2.2.3.3   *Kythnos (Siemens with Siemens 120 Wp modules)*

A 100 kWp PV array augments the power available from the existing diesel power station and an experimental wind park on the Greek island of Kythnos. It saves costly diesel fuel and enhances the reliability of the supply to the islanders. The battery capacity is 630 kWh.

2.2.3.4   *Rondulinu (Leray-Somer with France-Photon 36 Wp modules)*

A 44 kWp stand-alone plant supplies AC power for lighting, water pumping and refrigeration to a small hamlet in the mountains of Corsica. The battery capacity is 430 kWh. A 25 kVA diesel generator provides backup and can work in parallel with the PV system.

2.2.3.5   *Vulcano (ENEL with Ansaldo 33 Wp and Pragma 55 Wp modules)*

This 80 kWp system, with a 550 kWh battery, on a volcanic island north of Sicily, can either feed power into the island's grid in parallel

Figure 2.1:   The 80 kWp photovoltaic array at Vulcano, Italy. (Photo courtesy: Franco Roco and Salvatore Castello, ENEA.)

with the existing diesel station or supply a dedicated load of about 30 isolated homes.

2.2.3.6 *Mont Bouquet (Photowatt with Photowatt 71 Wp modules)*

A 50 kWp array with a 176 kWh battery, installed on the summit of Mont Bouquet, near Nimes in Southern France, partially powers three radio and three television transmitters with a total power consumption of between 35 and 40 kVA. Grid backup is provided.

2.2.3.7 *Pellworm (AEG-Telefunken with AEG-Telefunken 19.2 Wp modules) (Fig. 2.2)*

This 300 kWp system with a 2770 kWh battery, the largest and most northerly of the pilot plants, supplies all the electrical requirements of an enclosed recreation center on Pellworm, a German island in

Figure 2.2: The 300 kWp photovoltaic array on Pellworm Island, Germany. (Photo courtesy: Editor's collection, given to him by F.C. Treble.)

the North Sea. The loads consist of stoves and other items in the restaurant, heat pumps and water pumps, the indoor swimming pool, lighting, TV, radio, etc. As the center is used mostly in the summer, the load is well matched to the solar input. Any surplus energy is fed to the local grid.

2.2.3.8   *Chevetogne (I.D.E. with I.D.E. 33 Wp modules)*

A 63 kWp array, set on a grassy hillside in a large camping and sports center near Rochefort, Belgium, provides 40 kWp for the pumps of a solar-heated outdoor swimming pool and 23 kWp for lighting and ancillary services. The battery capacity is 275 kWh. Surplus power is fed to the grid, which also provides backup.

2.2.3.9   *Terschelling (Holec with AEG-Telefunken 19.2 Wp modules)*

This PV/wind hybrid system, consisting of a 50 kWp array, a 30 kW wind generator and a 180 kWh battery, provides 95% of the electrical power needed by a marine training school on Terschelling, a large island off the Dutch coast. The system is integrated with the local grid, into which excess power may be fed.

2.2.3.10   *Fota (University College, Cork, Ireland, with AEG-Telfunken 19.2 Wp modules) (Fig. 2.3)*

A 50 kWp array and a 160 kWh battery supply AC power to a dairy farm on an island south of Cork. The loads consist of milking machines, coolers and washing apparatus. The demand, peaking at 15 kW, is well matched to the solar input, as, in Southern Ireland, about 16 times more milk is produced in the summer than in the winter. Excess power is fed into the grid, which also provides backup. The space below the single-plane array is divided into bays, two of which serve as the Control and Battery Rooms and the others as cowsheds.

Figure 2.3: The 50 kWp photovoltaic roof on Fota Island, Eire. (Photo courtesy: Jeanine and Noel Cotter, Luminalt.)

2.2.3.11 *Zambelli (Pragma with Solaria 63.5 Wp modules)*

A 70 kWp array with a 27 kWh battery provides power to pump water from an existing tank on a mountain above Verona in Italy to another tank 350 m further up, as part of the water supply to the district. Two commercial piston pumps, each capable of maintaining a flow of 30 m^3/h, are driven by 35 kW AC motors supplied through a variable-frequency inverter.

2.2.3.12 *Tremiti (Italenergie with Siemens 120 Wp and Ansaldo 33 Wp modules)*

This 65 kWp stand-alone system with a 500 kWh battery, installed on a cliff top on a small island in the Adriatic, is designed to desalinate seawater by the reverse osmosis process, producing about 4000 m of potable water per year. The average energy requirement is $13 \, kWh/m^3$. The high-pressure, variable-flow pumps are driven by AC induction motors, which are powered by separate variable-frequency inverters. At present, water for the residents and summer tourists has to be imported by tanker from the mainland. There is no backup generator.

2.2.3.13 *Giglio (Pragma with Pragma 55 Wp modules)*

Installed on an island off the west coast of Italy, this plant comprises two systems — a 30 kWp array and 60 kWh battery for a commercial cold store and a 15 kWp array with a 200 kWh battery for a water disinfection plant (ozonizer). A novel feature of the cold tore system, which includes a five-cylinder compressor, is that the motor speed and the number of compressor cylinders in operation are automatically controlled to match the available PV power. The ozonizer output averages $150 \, m^3/day$ at present but it is planned to increase this. Grid backup is provided.

2.2.3.14 *Hoboken (E.N.I. with I.D.E. 33 Wp modules)*

A 30 kWp array, installed on the roof of a warehouse in a metallurgical factory near Antwerp, provides DC power for the production of hydrogen by the electrolysis of water. There is no battery. The hydrogen, stored at a pressure of 6 bars in four 1600 liter storage tanks, is used for metallurgical processing. Excess power is used for water pumping. Grid backup is provided through a 15 kW rectifier.

2.2.3.15 *Nice (Photowatt with Photowatt 71 Wp modules)*

A 50 kWp array, mounted on the roof of a building at Nice International Airport, supplies power for electronic equipment in the control tower and the sound system in the Main Hall. The battery capacity is 176 kWh. The sophisticated electronics require a continuous 5 kVA supply, free of distortion and micro-cuts — a requirement which can be admirably met by photovoltaics. Grid backup is provided.

2.2.3.16 *Marchwood (BP Solar with BP Solar 33 Wp modules)*

A 30 kWp system with an 88 kWh battery, installed near a power station on the south coast of England, near Southampton, can be operated in stand-alone or grid-connected modes, with or without the battery. When operated in stand-alone mode, it supplies selected individual loads.

In addition to the above plants, the Commission sponsored an experimental installation by ENEL of four 2.5 kWp PV arrays at Adrano, Sicily, on a site adjacent to Eurelios, the European 1 MW solar thermodynamic power plant. Two of the arrays are mounted on MAN dual-axis sun trackers, while the others are fixed. The aim is to compare different modules and evaluate the energy gain from two-axis tracking.

2.2.4 *Acceptance tests*

Between December 1982 and September 1984, a CEC team consisting of experts from JRC (under the leadership of Dr. K.-H. Krebs, with the author as consultant) visited each plant as soon as possible after completion and carried out official acceptance tests [5]. The tests consisted of a visual inspection of the plant, functional tests and on-site determination of the rated power of the array and the internal losses from module mismatch, cabling and diodes. Performance tests and functional checks were waived at Kaw because of the remoteness of the site and the similarity of the plant to that at Aghia Roumeli.

The accurate determination of the rated power, i.e., the net maximum output power from the array under standard test conditions (AM 1.5, 1000 W m^{-2} and 25°C) was particularly important, as the Commission's contribution to the cost was based on this parameter, as measured on site. JRC developed a portable test equipment for the purpose, which worked very satisfactorily. The on-site measurements proved to be in good agreement with the laboratory performance tests on the modules at Ispra.

2.2.5 *Monitoring*

Every participant in the program was required to record hourly values of the following parameters, derived (preferably) from measurements taken at 1 minute intervals or from instantaneous samples taken every 10 minutes:

- Global irradiation
- Irradiation at the array tilt angle
- Wind speed and direction
- Ambient temperature
- Mean module temperature
- Array output energy
- Converter/MPPT output energy*
- Battery input energy*
- Battery output energy*
- Inverter input energy*
- Inverter output energy*
- Energy to (+) and from (−) the grid*
- Energy from the backup generator*
- Energy to each load
- Timing of system switching events

(*where applicable)

The recordings were required to be made in a standard format on tape or disc and sent every month to JRC for processing and analysis. Finally, a log book was to be kept to record failures and accidents (with remedial action), severe weather, maintenance operations, sensor calibrations, any changes in array orientation, etc. [6].

Unfortunately, these requirements were not formulated until after the contracts were placed and, as a result of this, together with technical and organizational problems, the monitoring has been far from complete. Full data have been recorded for only two of the plants (Fota, Marchwood), while interrupted recordings have been received from ten (Aghia Roumeli, Chevetogne, Giglio, Kaw, Kythnos, Pellworm, Rondulinu, Terschelling, Tremiti and Vulcano).

No data have been received from Hoboken, because of problems with the data acquisition system and the hydrolyzer, nor from Mont Bouquet and Nice, because of fire damage. However, since 1985, the situation has improved. Results are being kept under review by the European Working Group on Photovoltaic Plant Monitoring, a body of JRC experts, contractors' representatives and consultants which meets every 6 months under the chairmanship of Dr. Krebs.

The information which follows has been obtained from the results of the acceptance tests, the monitoring records and a CEO-commissioned inspection of 11 of the plants, which took place between October 1985 and January 1986.

2.2.6 *Modules*

All the modules embody 100 mm diameter single-crystal silicon solar cells, except two types, which incorporate $100 \, mm^2$ multicrystalline silicon cells. The following encapsulation systems are represented:

	Window	**Encapsulant**	**Backing**
1.	Glass	Transparent silicone	White silicone
2.	Glass	Transparent silicone	Glass
3.	Glass	Polyvinylbutyral (PVB)	Glass
4.	Glass	Ethylene vinyl acetate (EVA)	Glass
5.	Glass	PVB	Tedlar-coated foil
6.	Glass	EVA	Tedlar-coated foil

Performance measurements at JRC and at the acceptance tests showed that, in many cases, the average rated power of the modules was less than the manufacturer's nominal figure. In the worst case, the discrepancy was 14.6%. Average module efficiencies, based on the total surface area, ranged from 6.5% to 8.5%. This is low by modern standards but one must remember that these were "first generation" European modules. Up to the time of the second inspection in 1985/6, about 0.25% of the modules had been renewed, mostly for cracked windows, which did not affect the electrical performance.

This damage has been attributed to a severe hailstorm (Giglio), vandalism (Terschelling) and design faults (Fota, Terschelling and Pellworm). The modules used in the latter three plants are clamped to the structure by washered bolts, which tend to loosen after some time, causing slippage. At Fota, there is insufficient clearance between the modules and the supporting galvanized steel beam, with the result that the glass is sometimes over-stressed when the mounting bolts are tightened. Another poor feature of these modules is the thin pigtails used for interconnection, which are easily damaged and difficult to stow neatly. The manufacturer has corrected these faults in later designs.

Some of the modules at Hoboken and Chevetogne showed signs of delamination and moisture ingress but this does not appear to have significantly affected performance as yet. Improvements in laminating techniques should make this type of defect a thing of the past.

The controversy over the optimum size of modules has not yet been resolved. Small modules (Fota, Terschelling and Pellworm) are cheaper to replace and may have advantages in manufacture and testing but large ones (Kythnos and Tremiti) yield a higher module efficiency, need simpler structures and require less on-site labor. Pre-assembled panels of small modules (Rondulinu) may prove to be the best answer.

2.2.7 *Array design*

In all except two cases, the rated power of the array, as measured on-site with an estimated ±5% margin of error, was within 4% of the nominal value. Array efficiencies, based on total subarray area, ranged from 5.1% to 7.8%. In no case did the array losses exceed 5% of the nominal rated power.

All the arrays fit in well with their surroundings thanks to the involvement of architects in the design process — a requirement of the Commission. In many cases, however, this has meant a considerable amount of ground clearance and site preparation. Uneven sites have had to be flattened, and sloping ones terraced and supported by retaining walls. Afterwards, the disturbed earth had to be planted or otherwise covered (at Vulcano with concrete) to keep the dust down. The projects with the lowest site preparation costs have been those on existing structures, e.g., roof at Hoboken or hard-standing at Marchwood, or cases where it has been possible to leave the contours and ground cover undisturbed (Chevetogne and Pellworm). Finding a second use for the space below the arrays, for example, the cow sheds at Fota and the sheep grazing at Pellworm, will also help to keep costs down.

The array support structures, in a variety of designs, have given no trouble, although some are over-designed, with massive concrete foundations cast *in situ*. Most designers opted for galvanized steel, but aluminum was used in two cases and, at Pellworm, half the array field is supported on hardwood frames. These have served well but wood shrinks with age and, consequently, the module clamping bolts have to be tightened once a year.

Galvanized steel fastening screws proved to be a problem at Hoboken, where heavy corrosion led to an expensive renewal campaign.

At Fota, where the array is in a single plane, a "cherry-picker" platform and special tools are required when one wants to replace one of the higher modules. This could have been avoided by the provision of laddered gangways between sections of the array, as at Hoboken. Elsewhere, no problem has been encountered in module replacement.

There appears to be no consensus, at present, on how to protect the array field against lightning. At Kythnos, ten tall steel masts were erected at intervals over the array field, while at Giglio, three single-span cables were suspended over the arrays from masts positioned outside the field. In all other plants, reliance has been placed on a good earthing system, with varistors to conduct lightning-induced currents to ground. To date, the only reports of lightning damage have come from Kythnos (data monitoring system) and Vulcano (one module). But at Mont Bouquet and Nice, the varistors proved to be incapable of withstanding the maximum system voltages under normal operation. Excessive leakage currents caused progressive degradation of the varistors, and in July 1983, they failed catastrophically at both plants and caused fires, which damaged the power conditioning equipment. After repair, in 1986, Mont Bouquet had the misfortune of suffering another fire (this time from the surrounding woodland), which destroyed much of the array field.

At several sites, disturbances occurred in data acquisition and control systems, which were attributed to inadequate shielding of the DC and AC power lines.

2.2.8 *Inverters*

With the sole exception of the hydrogen plant at Hoboken, all the pilot plants employ solid-state inverters to convert some or all of the generated DC to AC. However, some designers opted for commercially available inverters, with minor modifications and filter switching to improve the low-load efficiency, while others chose new designs embodying high-power transistors in the quest for better performance. Again, some decided on single inverters, while others sought higher overall efficiency by using multiple units switched in and out to suit the load. Instantaneous inverter efficiencies range from 90% to 96% at full load, depending on the design and the acceptable degree of harmonic distortion in the output waveform. But at 10% load, some efficiencies drop below 80%. Since the inverters are working for most of the time under less than full load conditions, the energy loss over a period can be considerable. This loss is made worse by the energy consumed by the inverter on zero load, unless the device is automatically switched off when not in use. The energy efficiencies of the inverters have been found, by monitoring, to vary between 60% and 85%.

Inverter malfunction has been one of the principal causes of plant shut-down. The main problems are electronic faults, synchronization faults, high starting currents, poor frequency control (in variable frequency types) and sensitivity to high temperatures, dust and humidity. These troubles have been compounded in some cases by the shortages of spare parts and skilled maintenance personnel.

2.2.9 *Maximum power point trackers*

A minority of the systems embody a maximum power point tracker (MPPT), a special type of DC–DC converter in which the voltage ratio is varied electronically to keep the array operating at or near its maximum power point under all conditions of irradiance, temperature and load. Giglio and Marchwood use single units but Tremiti has three and Kythnos has four. At Kythnos, the MPPTs are also used as part of the control system to reduce the PV energy flow when necessary, without switching out sections of the array. At Terschelling, there are no less than 29 MPPTs — one for each of the subarrays. The argument for this complexity is that the subarrays do not have to be matched and the system can accommodate partial shadowing and differing rates of degradation.

The proponents claim that MPPTs give a significant net energy gain but the opponents say that these devices are not yet fully developed and that the small energy bonus is not worth the added expense and complication. Experience has given some support to the latter view. At Kythnos, the power supplies to two of the MPPTs were damaged, and at Terschelling, the MPPTs have proved to be unreliable, mainly due to transistor defects. However, no quantitative assessment of the net gain or loss can be made until the data have been more fully analyzed.

2.2.10 *Power management and control*

All the plants have means for automatically controlling the flow of energy from the array to the loads, to and from the batteries (where fitted), and from the grid or backup generator (where applicable), in accordance with load requirements and the prevailing conditions of irradiance, temperature and battery state-of-charge. This is a difficult problem, made even more difficult by the fact that there is, at present, no really satisfactory means of monitoring the state-of-charge. Estimates made from battery current, voltage and temperature, as used in some battery charge controllers, are not reliable, as the relationship is affected by aging.

Six of the plants (Aghia Roumeli, Kaw, Mont Bouquet, Nice, Rondulinu and Vulcano) have simple control systems. These systems are reliable, as there is less to go wrong, but they have the disadvantage of relatively poor battery management.

Chevetogne, Fota and Pellworm have more complex control systems embodying microprocessors. This results in better power management and the software can be improved in the light of experience. However, special expertise is needed for programming and fault detection and this has not always been available when required. Nevertheless, significant improvements have been made at Fota as a result of the continuous analysis of monitored data and the use of a sophisticated computer model.

At Giglio, Hoboken, Kythnos, Marchwood, Terschelling and Tremiti, the control system, while of similar complexity, is modular in design. This involves extra equipment but has the advantage that the control algorithms for each module can be used in other systems employing the same unit.

In eight of the plants, the designers opted for a proprietary battery charge controller called the "Logistronic". These units have not proved satisfactory; indeed, none is now operational.

In some plants (Giglio, Hoboken and Mont Bouquet), control is achieved by load matching. In other cases, when load requirements are being met and the battery is fully charged, the control system must limit the power available, to avoid overcharging the battery. This is done by feeding surplus power to the grid (Chevetogne, Fota, Marchwood, Pellworm and Terschelling), by disconnecting the entire array (Aghia Roumeli, Kaw and Rondulinu) and/or by disconnecting the requisite number of subarrays (Fota, Nice, Pellworm, Terschelling, Tremiti and Vulcano). At Rondulinu, battery overcharge protection is effected by switching in a resistor network to move the operating point of the array.

2.2.11 *Batteries*

At Tremiti, Fota, Giglio and Vulcano, the battery has proved to be under-sized, thus aggravating the problem of overcharge protection.

At Fota, Pellworm and Terschelling, the battery bank is divided into two parts, each of which can be connected to either the main bus or an auxiliary bus. This arrangement gives great flexibility and enables one battery to be fully charged over a long period without discharge interruption, while the other supplies the load during the night and periods of low irradiance. It was expected to extend the life of the batteries — an assumption yet to be confirmed. However, computer simulation studies at Fota have indicated that the contribution of the plant to the dairy farm load would be increased by over 60% by combining the two battery banks and changing to a single bus system. This change would also increase reliability by reducing the amount of switching and simplifying the software.

The most common battery problem so far has been excessive drain due to shortcomings in the charge-control equipment. Probably because of this, 12 cells have failed at Aghia Roumeli and a further eight at Kaw. These are the only battery failures reported to date. Battery energy efficiency has proved to be lower than that predicted from manufacturer's data.

2.2.12 *Loads*

The matching of the PV system performance with the load has caused problems at most of the plants. In two cases (Aghia Roumeli and Pellworm), difficulties arose due to the uncontrolled addition of loads after commissioning. At Terschelling, the load was under-estimated. At Chevetogne and Rondulinu, the loads turned out to be less than anticipated. However, only minor problems have been encountered in the cases where the load characteristics were well-known or could be controlled by the operator of the plant. At Hoboken, the catalyzer and membranes in the electrolyzer have given trouble. The ozonizer at Giglio and the membranes in the reverse osmosis unit at Tremiti have also proved to be problem areas. The variable nature of the PV supply may be partly to blame for these defects.

2.2.13 *Operation and maintenance*

In general, the pilot plants have required minimal maintenance. At none of the sites has manual cleaning of the modules proved necessary, as deposits of dust and dirt are cleared away periodically by rain. Losses from contaminated windows are estimated to be generally less than 5%, although at Hoboken, a very dirty site, they have amounted to 10%. The textured glass surface on some modules tends to increase dirt retention and is a feature to be avoided in future.

Batteries have so far not needed to be topped up very frequently. The maximum frequency is twice a year (Terschelling) and the minimum frequency is once every two years (Pellworm).

Fota, Marchwood and Pellworm are fully automatic, needing no operators. At the other end of the scale, the desalination plant at Trelliti requires one full-time operator and three extra staff in the summer.

2.2.14   *Plant performance*

The mean monthly efficiency of the PV array, defined as the ratio of the monthly array output energy and the product of the array area and the monthly irradiation in the plane of the array, has been shown by the monitored data [7] to range from 0.95% to 7.39%. The poor performance at the lower extreme can be attributed, *inter alia*, to poor load matching, inadequate battery capacity, plant malfunctions and non-optimized control. Of course, the performance would be improved all round by the more efficient modules of the best type now available (12.9%).

The mean capacity factor, defined as the ratio of the monthly total array output energy (kWh) to the product of the nominal rated power (kWp) and the number of hours in the month, ranges from just over 5% for Marchwood and Fota (taken over a full year) to 10% at Vulcano (taken over the period July to November 1985). In their best months, Vulcano has a capacity factor of 20% and Kythnos 17%. Data from other plants has so far not been complete enough to compute comparable figures.

2.2.15   *Future program*

Recognizing that the 16 pilot plants constitute a unique tool for the further development of PV system technology, the Commission

(DG XII) has invited all the contractors to submit proposals for cost-sharing contracts to improve their plants in the light of experience. The response has been very encouraging and, at the time of writing this paragraph (December 1987), the Commission is in the process of finalizing the contracts. Examples of the improvements contemplated include the following:

1. Improve energy management and battery charge control (Aghia Roumeli, Fota, Kaw, Pellworm, Terschelling and Vulcano).
2. Change from a two-battery to a single-battery system (Pellworm and Fota).
3. Increase the size of the array (Giglio and Tremiti).
4. Combine the two array subfields (Giglio).
5. Increase the battery and water storage capacity (Giglio and Tremiti).
6. Increase the load to match the PV generator capacity (Chevetogne, Fota and Tremiti).
7. Install the automatic load management system (Chevetogne).
8. Improve MPPTs and array cabling (Terschelling).
9. Locate and replace inefficient components (Tremiti).
10. Make plant operation automatic (Terschelling).
11. Reduce staffing requirements by automation (Tremiti).
12. Improve the data acquisition system (Aghia Roumeli, Giglio and Pellworm).
13. Move the plant to a remote site and operate as an automatic stand-alone system (Marchwood).

The involvement of the Greek authorities is being sought to ensure the continued operation and improvement of the plants at Aghia Roumeli and Kythnos. It is planned to re-commission the Mont Bouquet and Nice installations. The future of the Hoboken plant, where the owner has decided not to continue operation, is the subject of further negotiation.

In addition to the improvement of individual plants, the Commission is organizing concerted action by expert groups to study the following common problems in the context of the pilot plants, develop improvements and prepare guidelines for the future:

- Battery charge control and management
- Power conditioning and control
- Computer simulation and modelling
- Photovoltaic modules and arrays
- Data monitoring and evaluation
- Array support structures
- Social effects

2.2.16 *Conclusions*

Despite all the problems, there is no doubt that the CEC Photovoltaic Pilot Plant Program has made, and will continue to make, a major contribution to the development of photovoltaic systems and components.

The lessons learned so far may be summarized as follows:

1. If full value is to be obtained from the investment in a pilot plant, detailed arrangements should be made at the outset for the ownership, operation, maintenance, repair and monitoring of the plant over a specified number of years. These arrangements should be spelt out clearly in the contract.

2. A manual should be prepared for each plant, giving instructions for the operation, maintenance and repair of the plant and data monitoring system.

3. System reliability is all-important, particularly in remote plants. The design should therefore be kept as simple as possible even if this involves some loss of efficiency.

4. A single-battery and DC bus system is preferable to a dual system.

5. In designing the power management and control system, computer simulation studies based on the best available data should be carried out to determine the arrangement most likely to produce maximum system efficiency. The model should subsequently be updated as more realistic data become available.

6. The existing load and likely future changes should be assessed as accurately as possible with a view to optimum matching.

7. Adequate shielding should be installed to prevent interference with control and data acquisition systems.

8. In choosing the site and designing the array, special attention should be paid to the need to minimize the costs of transportation, ground clearance, site preparation and installation.

9. As a general rule, only components which have been design qualified in laboratory-type tests should be used in a pilot plant. This applies also to load appliances.

10. High-efficiency modules, now available on the market, should be used in future projects in the interests of lower balance-of-system costs.

11. Modules should be mounted by fixing holes to give positive location, rather than by clamping. Interconnecting cables should be properly supported and protected against damage by animals and birds. In large, single-plane arrays, provision should be made to facilitate the cleaning, inspection and replacement of modules.

12. More effort is needed to improve the low-load efficiency and reliability of inverters.

13. Maximum power point trackers should be used only when careful computer simulation studies have indicated that they will produce a worthwhile gain. These devices need to be made more reliable.

14. Further investigation is necessary to determine the best way to protect PV systems from lightning damage.

15. Steps should be taken to ensure that the varistors used to conduct lightning-induced currents to the ground are capable of withstanding, with a reasonable safety factor, the highest voltages they are likely to be subjected to in normal operation.

2.2.17 *References*

[1] K.-H. Krebs, H. Ossenbrink, E. Rossi, A. Frigo, F. and Redaelli, "Module qualification tests at ESTI". 5th EC photovoltaic solar energy conference, Kavouri (Athens), Greece, October 1983, pp. 597–603.

[2] W. Palz, (Ed.). "Photovoltaic power generation, Series C, Vol. 1, proceedings of the final design review meeting on EC photovoltaic pilot projects".

D. Reidel Publishing Co., Dordrecht, Holland, 1982.

[3] F.C. Treble, "The CEC photovoltaic pilot projects". 6th EC photovoltaic solar energy conference, London, UK, April 1985, pp. 474–480.

[4] F.C. Treble, "Power conditioning systems in the EC photovoltaic pilot plants". *Int. J. Solar Energy*, Vol. 2, 1983, pp. 63–86.

[5] G. Blaesser, K.-H. Krebs, H. Ossenbrink, and J. Verbaken, "On-Site acceptance testing of large photovoltaic arrays". 5th EC photovoltaic solar energy conference, Kavouri (Athens), Greece, Oct. 1983, pp. 592–596.

[6] F.C. Treble, "Monitoring of the CEC photovoltaic pilot projects". 5th EC photovoltaic solar energy conference, Kavouri (Athens), Greece, Oct. 1983, pp. 584–591.

[7] G. Blaesser, and K.-H. Krebs, "Summary of PV plant monitoring data, 1984–1985". 7th EC photovoltaic solar energy conference, Sevilla, Spain, Oct. 1986, pp. 84–88.

2.3   Discussion after Fred Treble's presentation

Roy: Can you tell us something about the costs of hardware and installation of these European systems?

Treble: In the case of the pilot plants, the costs were not monitored as these were not considered to be interesting parameters.

For the demonstration plants, the module costs were in the range of 7–14 ECU per peak watt. Of this the ECC contributed 7 ECU/Wp under the (optimistic) assumption that this would represent about half of the system cost. In the event, the total system costs turned out to be nearer to 21 ECU/Wp. (At that time, 1 ECU was similar to US$1.)

Faiman: What is the optimum strategy, based on the European experience, for interconnecting cells to form modules, and modules to form arrays?

Treble: The individual cells are usually connected in series to form a single string within the module. In the case of the large, 125 W, Siemens modules, several of these strings are connected in parallel.

These modules are then series connected in order to achieve the desired bus voltage and these large strings are then parallel connected in order to reach the nominal power rating.

Each module is fitted with a bypass diode in order to minimize the effects of shading and hot spots.

These strategies were adopted at the start of the ECC monitoring program, but no attempts have been made within this program to determine the extent to which this is the best way of doing things.

Q: What kind of concentrator systems were considered?

Treble: Only the ones having a relatively low concentration ratio — usually of the planar–parabolic type. One such system with a concentration ratio of about 25 was actually planned to give about 30 kW. Unfortunately, the reflecting laminate gave problems, the cells were non-uniformly illuminated, and the heat sinks did not function satisfactorily. In the middle of all these problems, the cell manufacturer ceased production, and all alternatives were found to be far too costly. In the end, only a single 1 kW unit was ever built and the whole project was scrapped. In any event, by using concentration in Europe, one would lose a large fraction of the total annual energy.

Tabor: Electric motor manufacturers have found that AC motors are more reliable than the DC variety. On the other hand, you have indicated that the DC/AC converters were major causes of problems on many of the European photovoltaic systems. If you were to design a photovoltaic water pumping system, which kind of motor would you use?

Treble: I would choose an AC motor but with one important qualification: I would use a high-frequency (i.e., higher than 50 Hz) inverter in order to increase the conversion efficiency but develop it to the point that it would give trouble-free field performance. The point is that there is no inherent technological reason why these devices should not be as reliable as any other components — there has simply not been the incentive to develop them hitherto.

Appelbaum: As a comment, perhaps I might add that even though AC motors were preferred in the past for high-speed pumping rates, today there exists the alternative of brushless DC motors.

Question: Does the ECC study program include the use of DC appliances?

Treble: A number of houses in the south of France and in Greece have photovoltaic power supplies, which operate DC appliances. The main problem, however, lies with the availability of the latter. It is a kind of chicken-and-egg problem: Manufacturers do not develop such appliances because they do not see market prospects and therefore potential purchasers are unable to create such a market. Perhaps this is one area where government spending could help.

Question: Do any of the ECC systems employ fixed-voltage inverters?

Treble: I do not think so. They are all designed to cope with a range of input voltages.

Kreider: Speaking as a solar-thermal man, I am surprised to learn that many of the problems that are being re-discovered with photovoltaic systems were solved during the 1970s when they first appeared on thermal systems. They could have been avoided the second time around!

Roy: It distresses me that there is no universally agreed-upon definition of system efficiency that would enable different systems to be compared.

Treble: I strongly agree. The ECC is currently in the process of defining a standard but, alas, it will only apply to photovoltaic systems.

Gilon: I do not think that a truly universal standard definition for efficiency would be very useful. For example, were you to refer everything to the total radiation incident on the collector aperture, then a high-concentration system might be operating perfectly and yet be giving zero efficiency at some specific moment. You might then, erroneously, conclude that a nearby, poorly operating, non-concentrating system was superior just because it happened to be giving, say, 3% efficiency at the same moment. What truly counts is the total useful power produced by the system and what its price turns out to be.

Question: What kind of environmental test standard is employed by the ECC program?

Treble: The "IEC '68" general specification for electrical equipment was based on the US and European versions of "MILSPEC". This has been the basis of the ECC "Specification 502", together with input from JPL's "Block 5" standard. However, we are at present in the process of revising "Specification 502" so as to take into account various aspects of the Australian, Japanese and Spanish standards.

Appelbaum: What is the meaning of the term "load matching" as you use it, and how are mismatch losses measured in the ECC program?

Treble: We measure the voltage–current characteristics of the individual modules and complete strings, and subtract the cable and diode losses. What remains is presumably the mismatch losses. These, incidentally, are usually small because the module manufacturers have developed computer programs that enable them to optimally arrange each batch of modules so as to minimize any mismatch losses.

Appelbaum: How do you monitor the state of charge of the batteries?

Treble: What one really needs to do is to keep the electrolyte well-stirred and to measure its specific gravity. This cannot be automated in a satisfactory manner.

Chapter 3

1991

3.1 Editor's Foreword

Vern Risser's keynote presentation at the *4th Sede Boqer Symposium* provided a modicum of balance at a meeting that was heavily dominated by solar-thermal presentations, as is evident in Volume 1 of this book. But it was also an important mirror for Fred Treble's review in Chapter 2, given 3 years earlier, of European PV systems. For, as is evident from both presentations, intensive work was already underway in the West to commercialize terrestrial PV power systems. This was in contrast, as will be seen in Chapter 4, to parallel work in the Soviet Union, where the emphasis was still heavily on the development of PV cells for use in various space applications.

3.2 Large PV systems — will they play a role in our energy future? (Dr. V. Vernon Risser)

A keynote lecture presented by Dr. V. Vernon Risser (Daystar, Inc., Las Cruces, New Mexico, USA)

3.2.1 *Introduction*

The decade of the 1980s was a good time to be involved with PV systems in the United States. In the early part of the decade, government programs were booming, money was available, and euphoric forecasts of gigawatt markets abounded. Based on the still remembered gas lines of 1973, and President Carter's push for energy independence, renewable energy research was popular and the PV budget reached US\$151 M in 1981. Some of this money was used to buy prototype PV systems, both flat-plate and concentrating. A program that envisioned grid-connected PV systems on every residential roof was initiated, and the proponents of amorphous silicon began to talk about the "thin-film revolution and US$0.50 per watt."

However, rosy predictions aside, the public would have to wait. The pendulum was already swinging back to a market-driven energy policy under the Reagan administration. Subsidies were scrutinized, tax credits dropped, and programs slashed for renewable technologies. The oil crisis was a thing of the past, we were told to feel good about ourselves, conserve energy only if we felt like it, and not to worry. The price of gasoline stabilized at about US$1.00 per gallon in the United States and consumers adjusted; renewable energy was a good thing but too expensive and too esoteric to concern the ordinary citizen. By 1987, the PV budget had dropped nearly 80% from its 1981 peak and most of the available funds were spent on materials research. The central station PV systems that were to dot the desert southwest were a mirage.

Still, progress was made. The quality of the PV product improved and large systems were designed and installed in record time. The versatility of PV power systems was demonstrated for systems of all sizes. Power conditioning systems and balance of systems equipment were developed and improved to a point where the promise of low-maintenance systems became a reality. Although the cost of PV systems was still too high for generation of bulk utility power, PV power began to win the economic comparison battle for a large number of remote applications. As the decade closed, the PV manufacturers that made flat-plate crystalline modules were selling all they could make — at a price 10–20% higher than it was in 1988. The government budgets are also increasing: US$46 million in 1991 and projected to be up 30% to around US$60 million in 1992. But as we start the 1990s, a large dose of realism is needed. The technology's proponents cannot dictate the market (or the price) for the product; that will come only when the demand increases. The most important thing for the industry is to develop a healthy industry. This will require an economically competitive technology, sustainable markets, and user acceptance. We should look at our past mistakes and learn from the experiences with the large systems installed in the 1980s. This paper will review the course of development of larger PV systems in the United States — from the 27 kW system installed in 1977 to the 200 kW systems installed at Davis, California, in 1991. Lessons learned are discussed and the current environment for large PV systems is assessed. Finally, the likely evolution of the use of PV systems future is predicted.

3.2.2   *The past: Policies and projects*

The Energy Research and Development Agency (ERDA) established the National Photovoltaic Program (NPP) in the United States in 1975. One important part of the program was to buy and evaluate the PV modules that were available in 1976, the so-called Block Buys conducted by the Jet Propulsion Laboratory (JPL). A second part of the NPP was the deployment of large (> 10 kW) demonstration PV systems. The systems were to be funded by the government and used to provide power for practical applications. Building and operating the systems would allow the performance and suitability of the systems and their durability to be ascertained.

3.2.2.1   *1977–1979*

The first four systems were installed between 1977 and 1979. See Table 3.1.

The notations used in this and subsequent tables are given in the following.

Table 3.1:   Flat-plate PV systems installed in the United States, 1977–1979.

Location	Date Installed	Size (kW)	Load	Module Type	Module Manufacturer	Status in 1991
Mead, NB	7–77	27.5	Pumps	FP/F	SX	Gone
Bryan, OH	8–79	15	Radio Station	FP/F	SX	Operating
Mt. Laguna	8–79	60	FAA Comm.	FP/F	SX,SP	Junk
NBNM, UT	12–79	100	Ranger Station	FP/F	SL, M, A/S,ST	Not Operating 53 kW

Module Type

FP/F	Flat Plate Fixed
FP/1A	Flat Plate Single-Axis Tracking
FP/2A	Flat Plate Two-Axis Tracking
FP/F/M	Flat Plate Fixed with Mirror Enhancement
LFT/T	Linear Focus Trough with Thermal Energy Used
PFF	Point Focus Fresnel
LFF	Linear Focus Fresnel
FP/CIS	Flat Plate Copper Indium Diselenide
FP/SOC	Flat Plate Thin Film Silicon on Ceramic
FP/CdTe	Flat Plate Cadmium Telluride

Module Manufacturers

SX	Solarex
SP	Solar Power Corporation
SL	Spectrolab
M	Motorola
A/S	Arco/Siemens
ST	Sensor Tech
SKI	Solar Kinetics Inc.
MS	Mobil Solar Energy Corporation
UPG	Utility Power Group

The modules used in the systems in Mead, Nebraska; Bryan, Ohio; Mt. Laguna, California; and Natural Bridges National Monument (NBNM), Utah (1977–1980), showed there was much to learn about module construction before a durable product capable of performing for 25 years was available. Many problems occurred [1, 2]. Many of the modules used the encapsulant/cookie sheet construction method with no glass cover. The dust and dirt accumulated on the soft encapsulant, and within a few years, the transmissivity to the cells had decreased by 30–40%. Other problems were cracked cells and poor cell-to-cell connections. However, the system at Bryan, Ohio, is still providing power to the radio station today [3]. The modules were replaced in 1981 with a more durable version and few array problems have occurred in the last 10 years. No records are being kept, but the radio station personnel believe that more than 75% of the power needs are being met with solar. They have made a commitment to keep the system operating and pay the cost of operation and maintenance required.

The system at NBNM is still capable of operating at a power level of about 50 kW. The Motorola modules have failed. The system at Mead became increasingly failure-prone and was dismantled. The Mt. Laguna system was tested and assessed in 1988. Multiple module failures had occurred and extensive work would have to be done to obtain any power. The system has not operated in years and the array is essentially junk. Most of the problems at Mead and Mt. Laguna were caused by either cracked cells or connections that failed. Both cell-to-cell failures and module connection failures occurred. The problems found in these early systems confirmed the need for the program at JPL to increase the durability and reliability of flat-plate modules. The program was conducted by buying and testing available modules and working closely with the module manufacturers to rectify problems. Blocks I, II, and III were completed in 1976, 1977, and 1978, respectively. Many problems were solved and tighter module construction specifications led to more durable modules for the systems installed in the early 1980s.

3.2.2.2 *1980–1982 flat-plate systems*

The next cluster of systems used Block 4 modules and was part of the PRDA-38 DOE procurement and the Residential Systems Program initiated in 1979, see Table 3.2.

The systems in Lovington, New Mexico; Beverly, Massachusetts; El Paso, Texas; and Oklahoma City, Oklahoma, began operation in 1981–82 [4]. The Bermuda system, built at the US Navy's Tudor Hill Laboratory, started operation in October 1983 and the San Bernardino, California, system began operation in early 1982.

These systems were not part of the PRDA activity *per se* but were funded by the government through different channels. The systems at El Paso, Bermuda, and Oklahoma City are still operating. San Bernardino has been dismantled and removed. The systems at

Table 3.2:   Flat-plate PV systems installed in the United States, 1980–1982.

Location	Date Installed	Size (kW)	Load	Module Type	Module Manufacturer	Status in 1991
El Paso, TX	1–81	17	Power UPS	FP/F	SP	Operating 13 kW
Lovington, NM	3–81	95	Shopping Center	FP/F	SP	Not Operating
Beverly, MA	9–81	90	High School	FP/F	SP	Not Operating
Oklahoma	2–82	90	Museum	FP/F/M	SX	Operating
Bermuda	8–83	45	Navy Buildings	FP/F	A/S	Operating 42 kW
San Bernardino	1982	35	Light Industry	FP/F	SX	Gone

Lovington and Beverly are not operating because of problems with the power conditioning systems (PCS).

The El Paso system was installed at the Newman Power Station operated by El Paso Electric Company. The power station maintains an uninterruptible power supply (UPS) to provide a controlled shutdown sequence in case of system power failure. The PV system was interconnected with the existing UPS. The array was originally rated at 17.5 kW DC and was rated at 14 kW in 1988. The drop in power was caused by some connection failures in the module junction boxes and significant soiling of the modules caused by the precipitate from nearby cooling towers. Little maintenance has been performed on the array. The controls have been removed from the system so the array provides power to the large UPS battery bank any time the sun is shining. The system at Oklahoma City is installed on the roof of the Omniplex museum and interconnected with the local utility. The array uses mirrors to enhance the output of Solarex flat-plate modules. The enhancement never reached its design potential. The modules are installed in single rows with the modules lengthwise. The modules are tilted at 45° and facing each row is a 45° tilted mirror. The mirrors did not provide uniform illumination and the increased irradiance on some cells caused uneven performance and higher mismatch losses. As a result, the array never produced as much power as expected. (Uneven temperature distributions contributed to the problems.) Still, the system is producing today and the Omnion PCS has probably the best availability record of any unit. The PCS used in the Bermuda system was updated in 1989 and that system continues to provide power to the Navy facilities on the island. There has been no measurable degradation in array performance.

The Lovington, New Mexico, and Beverly, Massachusetts, systems were designed as sisters with identical components and configuration. The intent was to study the performance difference of identical systems installed in different areas. Neither system has been operated consistently since 1985–1986 because of the erratic performance of the power conditioning systems. As reported earlier, many of the problems with the PCS were not hard failures [5]. This era of PCS development included a large amount of circuitry to monitor conditions of array and utility status. Detected conditions outside preset tolerances caused the PCS to shut down. Some of the conditions would allow an automatic restart; most required a manual restart. These "nuisance" shut-downs proved a bane to site operators. Some reported visiting the sites "when time allowed" and trying to restart the PCS. If they were not successful, they would try again the next time they happened to be on site. Often, the PCS would fail to restart several times (over a period of weeks) and subsequently start operation on the fourth or fifth attempt. No corrective action had been taken. This exasperating behavior of the PCS soured site operators on PV systems and placed a cloud over claims of low O&M costs.

The San Bernardino system was installed on the roof of a manufacturing plant to demonstrate that PV power could be used as an augmentation for light industry. The system performed as expected but the program was ill-conceived with little support from the manufacturing plant operator. When problems occurred, they were not fixed, so the system was removed. The reputation of solar power was not enhanced. There were two other projects that fit in this same category. A 240 kW system using linear focus troughs was installed at the Mississippi County (Arkansas) Community College in 1981. At the Northwest Mississippi Junior College, three 50 kW

Table 3.3:   Concentrator PV systems installed in the United States, 1980–1982.

Location	Date Installed	Size (kW)	Load	Module Type	Module Manufacturer	Status in 1991
Albuquerque, NM	1982	47	Office Building	LFT/T	SKI	Gone
Dallas, TX	6–82	25	Airport/ Hotel	LFF/T	Entech	Not Operating
Kauai, HI	1982	35	Hospital	LFT/T	Accurex	Junk
Phoenix, AZ	3–82	225	Airport	PFF	Intersol	Gone

systems were to be operational in 1981. Both these projects were funded by direct grants from the government and the systems' performance never reached advertised goals. The electricity produced never exceeded the amount required to install the hardware.

3.2.2.3   *1980–1982 concentrator systems*

During the same time period that the PRDA-38 projects were being designed and installed, the NPP was sponsoring the PRDA-35 development of concentrating systems. Four systems were installed in 1982–1983 as part of this effort, see Table 3.3.

Two of the four, at Kauai, Hawaii, and Albuquerque, New Mexico, used linear focus troughs (LFTs). The largest system of the group, installed at the Phoenix, Arizona, airport, used point focus Fresnel (PFF) lens and the system at the Dallas, Texas, airport used linear focus Fresnel (LFF) lens to concentrate the sunlight. All the systems except the one in Phoenix planned to use the heat from the active cooling system to preheat water for industrial use: in Kauai for a hospital, in Dallas for a hotel, and in Albuquerque for an office building.

None of these concentrator systems are operational. The system in Dallas, installed by Entech, Inc., operated until 1989 when a building modification damaged the tracking controllers. The modules are still installed and some cursory testing indicates that the performance has not degraded significantly. The system at the Phoenix airport has also been dismantled. The city of Phoenix wanted the land for other uses, so the modules were removed and distributed to universities where they are being used in test programs. Some of these modules had been damaged by water and the power rating on this system had dropped by about 40% from the original value. The linear focus trough systems in Kauai and Albuquerque never operated successfully. They were sensitive to the slightest misalignment and required a jot of "hand tuning" to get the power output near predicted. It is doubtful if the electricity produced ever exceeded the O&M labor costs. Also, the system in Kauai was operated without the cooling system activated and the high temperatures charred some cells and turned others dark brown. When the system was tested in 1985, its rating was estimated at 7 kW. This system was moved from the hospital near Lihue, Kauai, to the US Navy base on the west side of Kauai in 1985. The system was re-installed but never operated electrically. Some water was heated for use at the base barracks. The system in Albuquerque is still on the roof of the BDM Corporation building in Albuquerque but has not operated for years.

3.2.2.4 *1983–1990*

The government spending for PV system development almost dried up in the mid-1980s. Even the money required to monitor the performance of the installed systems was hard to obtain. However, more large systems were installed through private or cooperative funding ventures. Arco Solar Corporation spearheaded this drive by installing the first megawatt size system in Hesperia, California. This system, actually about 750 kW, began consistent operation in 1983. Arco Solar followed Hesperia with the 5.2 MW system installed on the Carrisa Plains in central California in 1983–1985. Table 3.4 gives a list of 11 large systems installed in the mid- to late 1980s, excluding those in the Photovoltaic for Utility Scale Applications (PVUSA) project. The PVUSA systems will be discussed separately.

The systems at Sacramento Municipal Utility District (SMUD) and Georgetown University were funded primarily by DOE. The Puerto Rico system was funded by the island's utility company. The systems at Hesperia, Carrisa Plains, and John Long (Phoenix) were funded privately. The PV300 system in Austin, Texas, was a joint

Table 3.4: PV systems installed in the United States 1983–1990.

Location	Date Installed	Size (kW)	Load	Module Type	Module Manufacturer	Status in 1991
SMUD1	8–84	1000	Bulk Power	FP/1A	A/S	Operating
Georgetown	9–84	250	University Building	FP/F		Operating
Hesperia	8–83	750	Bulk Power	FP/2A	A/S	80%Operating
SMUD2	7–85	1000	Bulk Power	FP/1A	A/S, SX, MS	Operating
Carrisa	1986	5000	Bulk Power	FP/2A	A/S	Operating
John Long	8–85	170	Residences	FP/F	A/S	Operating
Austin PV300	5–87	285	Bulk Power	FP/1A	A/S	Operating
Puerto Rico	1–87	85	Bulk Power	FP/F	SX	Operating
Alabama Power	1987	75	Demonstration	FP/F	Chronar	40%Operating
Virginia Power	1–87	75	Demonstration	FP/F/ 1A/2A	SX,SX,A/S	Operating
Austin 3M	5–90	300	Bulk Power	LFF	Entech	Operating

effort between DOE and the Utility department of the City of Austin as was the 300 kW concentrator system 3M/Austin, which was the latest system installed. The systems at Alabama Power and Virginia Power were funded, in large part, by the utilities with some assistance from DOE. All of these systems are operating (at some capacity) today.

Two 1 MW flat-plate PV systems were installed at SMUD in 1984–1985. Single-axis powered tracking was used on both systems. A single PCS was used on each system: an Omnion on SMUD-1 and a Toshiba on SMUD-2. These were the largest inverters to be used to that date. In 1987, the PCS on SMUD-1 was destroyed in a fire caused by a ground fault [6]. The unit was replaced in 1989 with another Omnion unit and the system restarted operation. However, multiple ground faults in the underground subarray wiring kept the plant turned off much of the time. The wires used — locomotive cable — were evidently nicked in places during installation, and ground faults began to occur after 4–5 years. In 1990, SMUD replaced the underground wiring. In July 1991, the second highest monthly energy production on record was produced. There is no degradation that can be detected. The SMUD-2 array users suffered from some connection problems from the laminate to the busbar. This problem caused a few hundred modules to fail after several years in the field. The problem was fixed in 1990.

The systems at Hesperia and Carrisa were engineered and installed by Arco Solar Power Corporation (now Siemens Solar Industries). Both systems use two-axis tracking for the flat-plate modules. There are two inverters at Hesperia and 10 at Carrisa. All but one of the PCS is made by Delta Electronic Controls Corporation (DECC). The exception is a Toshiba PCS on one segment at Carrisa. Nine of the 10 subfields at Carrisa use mirror enhancement on the two-axis trackers. From 1983 to 1989, the Hesperia system was the most consistent power producer of any of the PV systems installed. With the exception of 1985–1986 when a major re-work of the gear drives took place, the system has produced about 2550 kWh for each kilowatt installed. The mirror-enhanced trackers at Carrisa caused the modules to operate at higher temperatures than expected.

After a couple of years of operation, some noticeable darkening of the encapsulant was noted. This browning has caused a significant decrease in module output, causing the rating of the system to drop below 3 MW [7]. The subfield without mirrors has not degraded. Arco Solar sold both Hesperia and Carrisa in 1989 and some modules have been removed from each site for resale.

The systems at the John Long residential development in Phoenix, Arizona, and the Austin PV300 system use Arco Solar laminates (unframed modules) similar to those installed at SMUD-2. The John Long array is fixed and the Austin system uses passive single-axis tracking. Both systems are operating well and have experienced few problems, although neither array attained its design ratings.

The Georgetown and Puerto Rico systems use Solarex Corporation modules. The Georgetown array is installed on the roof of a building and the modules are (almost) inaccessible for service or replacement. There are known failures in the array (about 50) but no modules have been replaced because of the difficulty in getting to the modules. The array in Puerto Rico has suffered only two connection problems. Both systems use Omnion inverters.

The Alabama Power array was the first large array to use amorphous silicon modules. Chronar Corporation supplied the modules that were installed in fixed arrays. Multiple module failures have occurred primarily because of shorting from the active thin film to the module frame. This was the first time that a-Si modules were operated at greater than 200 V DC and multiple shorts occurred, particularly when the modules were damp. The system is operating but the rating is down about 50% from the initial rating.

The systems installed at Virginia Electric Power Company's North Anna Power Plant were intended to demonstrate the effect of different mounting schemes. A fixed array, a single-axis tracking system, and a two-axis tracking system using flat-plate modules were installed. The systems are interconnected with the utility grid. Some reliability problems with the two-axis tracking unit have been experienced.

The most recent system installed was the 3M system in Austin, Texas. This system uses Entech, Inc., LFF concentrator modules similar to those used in the Dallas system installed in 1982. After some early problems were fixed, consistent energy production began in June 1990. A comparison of the performances of the two systems in Austin is included in Section 3.2.3.

3.2.3 *PVUSA*

The ongoing Photovoltaic for Utility Scale Applications (PVUSA) program was initiated in 1986 with the main purpose to give utilities experience with PV systems [8]. The PVUSA project is the only ongoing project that includes the installation of a number of large PV systems. This project is jointly funded by the USDOE, EPRI, the California Energy Commission, and the participating utilities. Pacific Gas & Electric is the principal utility sponsor, and the main testbed is located near Davis, California. Other utility companies are encouraged to join the project and share in the information and possibly host a test system in their service area. The first system in the program began operation in January 1989. Plans call for the installation of 12 systems during Phase 1 of the multi-year project: nine emerging technology (EMT) systems and three utility scalable (US) systems. As of August 1991, six EMT systems of 15–19 kW are operating, and three more are under construction. One US size system of 200 kW is operating, and acceptance tests are being conducted on the second. One 400-kW system is planned. See Table 3.5.

The performance of the EMT systems that have been installed is consistent with other systems installed in the late 1980s. The Arco/Siemens array using single-crystal silicon modules has shown a DC efficiency of 10.7% and an AC efficiency of 9.7% [8]. Only minor problems have occurred with the connections to two modules. Array performance is consistent and predictable and the peak rating is only 3% lower than proposed. Three of the EMT arrays use amorphous silicon modules: two systems use Sovonics (now United Solar Systems Corporation, USSC) modules and one uses modules made by Utility Power Group (UPG). One of the USSC systems is located at Maui, Hawaii, and operated by Hawaii/Maui Electric,

Table 3.5:   PVUSA systems.

Location	Date Installed	Size (kW)	Efficiency DC/AC	Module Type	Module Manufacturer	Status in 1991
Davis	1–89	18.7	10.7/9.7	FP/F	Siemens	Operating
Davis	6–89	17.3	3.2/2.8	FP/F	Sovonics	Operating
Maui	10–89	18.5	3.7/3.4	FP/F	Sovonics	Operating
Davis	12–89	15.7	3.2/2.9	FP/F	UPG	Operating
Davis	10–90	15.7	8.8/8.0	FP/F	Solarex	Operating
Davis	5–91	16.5	11.3/-	LFF/2A	Entech	Operating
Davis	8–91	174	- /11.3	FP/1A	Siemens	Operating
Davis	10–91	180	- /7.4	FP/1A	IPC/Mobil	Check Out
Davis	—–	400	- /3.4	FP/F	APS (Chronar)	Planned
Davis	12–91	19.4	5.5/-	FP/CIS	Siemens	Planned
Davis	12–91	19.3	7.4/-	FP/SOC	AstroPower	Planned
Davis	12–91	18.7	5.3/-	FP/CdTe	Photon Energy	Planned

a participating member of the PVUSA project. All of the a-Si arrays are operating consistently but their efficiency is in the 3.2–3.7% range. The last EMT system to be installed uses Entech LFF concentrating modules similar to those installed at the 3M/Austin system in 1990. This system started operation in June 1991 and has shown a peak DC efficiency of 11.3% while operating at 850 W/m^2 and 20° C ambient conditions [9].

The PVUSA project is providing two major benefits to the PV industry in the United States. First, it provides an incentive for pushing the newer technologies into production and creates a standardized testbed where the different types of systems can be evaluated and compared. The CIS, CdTe, and thin film polycrystalline silicon on ceramic systems will be the first of their kind installed on such a large scale. The lessons learned will likely be as beneficial as those learned from the systems installed in the early 1980s. Second, PVUSA is forcing the PV manufacturers to adhere to strict quality control regimen for their product. The introduction of the "wet Megger test" where the leakage current of the modules and the array source circuits is measured while a wetting agent is sprayed on the modules is the first of its kind. This test is indicative of the tough standards for module qualification that are being imposed. If 25-year systems are to be installed, these tests, like the JPL Block tests conducted earlier, will keep the emphasis on quality. However, the benefits that can be obtained from installing the larger US systems at PVUSA are more questionable. These systems are a little different from those installed in the late 1980s, and while the questions of product durability and array configuration optimization require study, there is a question whether the problems could not be addressed with 20 kW systems instead of 400 kW systems. It is especially difficult to see the need for a 400 kW amorphous system that will operate at 3% efficiency.

3.2.4   *System performance*

Data were obtained on a number of systems discussed in the previous section. Most of the flat-plate systems operated as predicted for the first few years. Various problems occurred, but maintenance was performed and so energy production was not interrupted for a long period. From the perspective of 1991, there are two important issues: consistent (dispatchable) production which depends on reliability of the system and its components, and array peak power ratings. If these systems were to be utilized in the future for bulk power production, the operators must know how much power to expect and they must be assured of consistent production. The data show a promise of consistent production but the method of rating arrays used by module manufacturers is not realistic.

3.2.5   *Energy production and reliability*

The consistency of annual production can be seen from the graph in Fig. 3.1, which was originally published in 1987 [5]. The normalized AC energy production values for four flat-plate systems and one concentrator system are given. The production figures are from meter readings on-site.

The slope of the production trend line for all the systems shows consistent production in the early years of the system's life. Then, for example, the production rate for the system in Lovington, New Mexico, (LCEC) fell significantly in 1986 because of PCS downtime. As mentioned above, the problems worsened in 1987, and the system

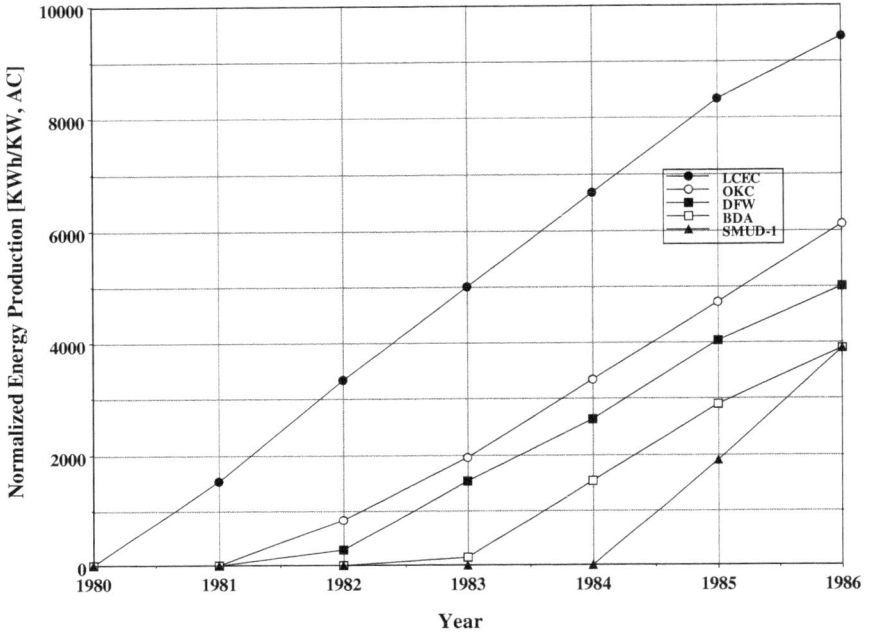

Figure 3.1: Energy production rates for five systems.

seldom operated after that date because of the large amount of time required to keep the inverters online. The concentrator system at Dallas-Ft. Worth Airport (DFW) also showed a drop in production in 1986 due to PCS problems. These proved to be temporary, and after the PCS was repaired the production rate resumed. However, in 1989, a building renovation damaged the tracking controller, and the system has not operated since. The other systems in Oklahoma (OKC), Bermuda (BDA), and Sacramento Municipal Utility District (SMUD PV-1) show no signs of production drop. However, SMUD-1 was down for nearly 2 years when the PCS was destroyed by a fire in 1987. Bermuda has been a consistent producer except for PCS problems that sometimes took the system off line for several weeks at a time. Oklahoma City has had the fewest problems that would put a drop in the production trend line, but data for that system have not been acquired since 1986. The system operator reports consistent system performance [10]. Similar plots

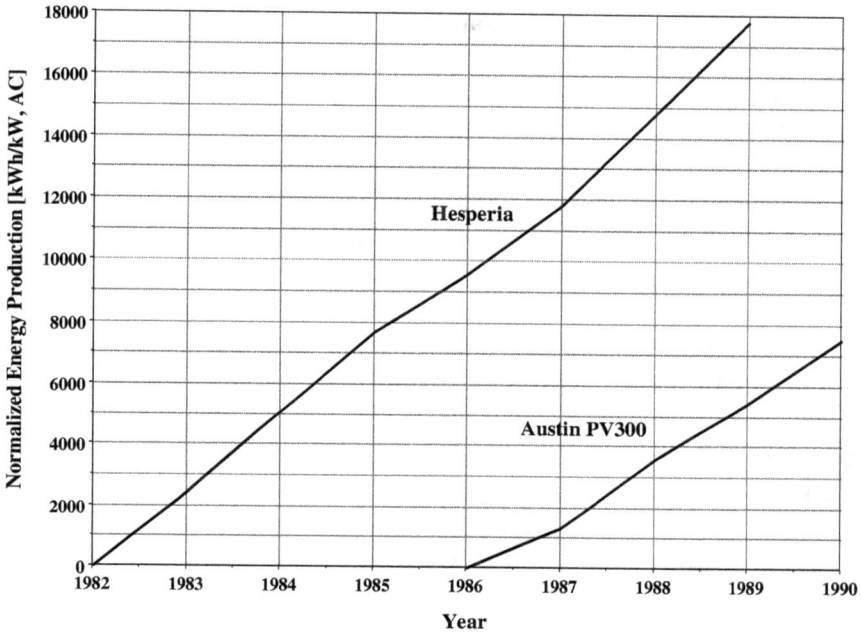

Figure 3.2:   Normalized production of Hesperia and PV300.

of energy production for different systems have been published in the Large System Evaluation Reports prepared by EPRI [11–14]. Unfortunately, there is no complete record for any system that is over 10 years old. The system in Bryan, Ohio, is the oldest one in continuous operation, although the modules were changed in 1981. For systems with data available, the ones at Hesperia and Austin PV300 have proven to be the most consistent operators. The production trends are shown in Fig. 3.2.

The production at Hesperia was affected by a sequential re-work of the gear drives in 1985–1986 but this two-axis tracking system maintained a slope of approximately 2550 kWh/kW installed until parts of the array were removed in 1990. The Austin PV300 system is a single-axis tracking system producing at a rate of approximately 2100 kWh/kW installed. The low production in 1987 was because the system was undergoing acceptance testing and did not start production until May of that year. No more data are expected from

Hesperia, which is being dismantled so the modules can be sold for remote system applications.

The factor that affects the annual production is the system reliability. Experience has shown that the reliability of the PCS is the most critical factor. It is strange that more attention is not paid to improving this critical factor. Most of the effort on PCS development is on improving the power electronics in the unit and obtaining one more percent improvement in efficiency. However, a 1 MW system in the southwestern US will produce about 2500 MWh of AC electricity or about 6.8 MWh/day. A 1% increase in PCS efficiency will yield 25 more MWh/year — if the PCS operates every day. Four lost days per year will offset the energy gained from the efficiency increase. Yet, a lot more activity is being expended on efficiency improvement than on long-term reliability improvement.

The module reliability for flat-plate modules is high. The number of failures internal to the laminate can be counted on one hand. However, there are other failures that can cause a good module not to contribute. The staff of the Southwest Technology Development Institute has tested many systems to determine what causes lost production. As reported in 1988, 13,468 of 265,000 modules were not producing when the arrays were tested [15]. The causes are summarized in Fig. 3.3.

Module reliability continues to be high, as only 0.2% of all the modules tested have "failed". (A failed module is any module with output reduced by 50% or more, which cannot be repaired without replacement.) Clearly, the largest cause of production loss in the DC portion of PV systems is the failure of balance of system (BOS) equipment. About 60% of the production loss caused by these BOS problems can be attributed to switch and fuse failures.

3.2.6 *System ratings*

Experience has shown that most manufacturers' module ratings do not accurately describe PV system performance. This is because manufacturers usually rate modules at standard conditions (1000 W/m^2 irradiance and 25°C cell temperature), seldom seen in the field. They will then calculate the price charged by multiplying

265,000 Modules Tested
13,468 Non-producing

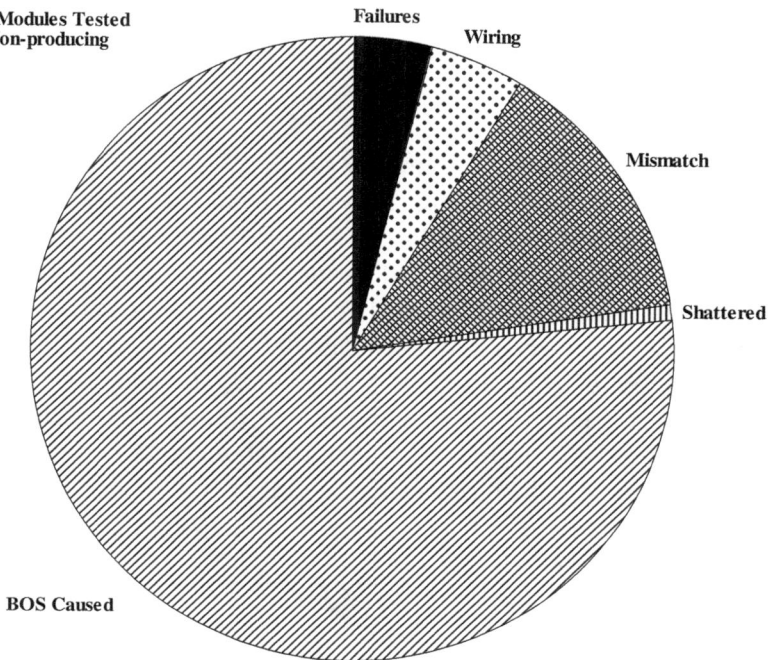

Failures

Wiring

Mismatch

Shattered

BOS Caused

Figure 3.3: Causes of non-producing modules.

the individual module rating by the number of modules in the array. This invoice rating should not be associated with predicted system performance. A closer estimate of expected energy production can be obtained by calculating the nameplate rating. This can be obtained by multiplying the invoice rating by a loss factor. The loss factor represents typical losses such as mismatch, soiling, wiring losses, and switchgear losses. While these losses are not constant from system to system, a good average for many systems can be obtained using the following factors:

PV Module/Panel Mismatch	0.2%
PV Panel/Subarray Mismatch	1.0%
Wire Loss	1.2%
Switch Gear Loss	0.2%
Blocking Diode Loss	1.1%

Soiling Loss   4.0%
Power Tracking Loss   0.6%
Combined Loss Factor   91.9%

Using this method, the nameplate rating for a system would be 91.9% of the invoiced amount of power. Note that these calculations are for DC power and no allowance for PCS efficiency is included.

A third rating, the installed rating, can be determined either by field measurement or by regressing long-term performance data. An accurate installed rating and the calculated nameplate rating should be nearly equal. The Southwest Technology Development Institute has measured many PV systems and determined the installed rating. In most cases, this is 10–15% below the nameplate rating and often 25–30% below the invoice rating. See Fig. 3.4.

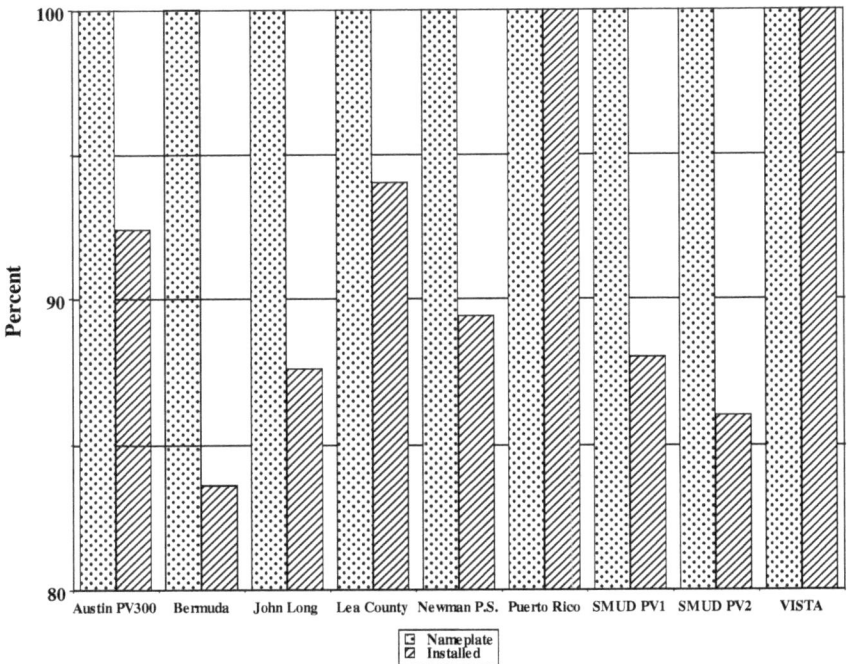

Figure 3.4:   Installed rating as percent of nameplate rating.

Even after losses are accounted for, the nameplate rating for all the systems except Puerto Rico and Virginia Power (VISTA) is higher than the installed rating. Since the nameplate rating is based on the manufacturers' ratings for their modules, one would have to believe the original ratings are not indicative of actual performance at the system level. Manufacturers' ratings have also been shown to be overoptimistic at the module level [16].

3.2.7 *Lessons learned*

What have we learned from more than a decade of designing, installing, and operating PV systems? Many things. The improvement in the durability of the basic flat-plate module has been dramatic. There are some questions about encapsulant discoloration — what effect it will have on module power production and how to prevent it from occurring — but the glass-covered, laminated module is nearly failure-free. Making durable connections from module to module requires better engineering on some modules and perhaps some added cost, but it is not an intractable problem.

Linear focus troughs will not work for PV systems because it is too difficult to maintain alignment and uniform illumination in tracking systems covering large areas. The potential cost benefits are illusionary because of the operation and maintenance costs involved with keeping this type of system optimized.

The annual operation and maintenance costs for flat-plate systems, either fixed- or single-axis tracking, can be held below $0.01/kWh produced. There is reason to believe this low value can be maintained over 25 years of system life if a reasonable preventive maintenance plan is implemented and spare parts are maintained on site for the PCS. Any system requires some maintenance, but a PV system will prove to cost less than most to keep it operating.

Two-axis tracking systems can be operated reliably but it is doubtful that the extra energy produced over a single-axis system will justify the added cost. For concentrator systems with high magnification ratios such as Alpha Solarco, Inc.'s system, the proof of reliable tracker operation is a cause for optimism.

You can't give systems away. There has to be a commitment
of interest and equity on the part of the operator to obtain long-
term performance. Unless there is an economic incentive to keep
the system operating, it won't. Too many of the systems have
been neglected after the first few years — often coinciding with a
discontinuation of government funds. Some demonstration systems
have paid great dividends because of the problems that have been
observed and subsequently solved. However, no one has seriously
considered building a PV plant to produce bulk power for the next
40 years — a planning process that is routine for other types of
generators. That is the biggest lesson of all — we have a long way to
go before these systems are economically feasible for bulk power.

3.2.8 *System cost*

The Electric Power Research Institute has made an assessment of
the cost of energy (COE) from potential large PV plants located
in the southwestern United States [17]. Using 1989 dollars, they
have studied what the module costs need to be for energy to be
sold at $0.08/kWh, a rate they reckon will be necessary before bulk
power plants are seriously considered. Their results, summarized
in Table 3.6, indicate the required module price for three different
efficiencies. The results are sobering for PV enthusiasts. The study
shows that a flat-plate module with 15% efficiency must cost only

Table 3.6: EPRI Estimate of the required module costs.

Module Type Southwestern U.S. System Site	Module Efficiency*	Module Costs (Dollars/m²)+
Fixed Flat Plate	10	27
	15	66
	20	105
Concentrator	15	26
	20	67
	25	109

Note: *25°C, 30 year life.
+Assumes additional BOS costs of $50/m² flat plate and
$100/m² concentrator.

$66/m^2$ or $0.44/W if it is to produce energy in the southwestern US for $0.08/kWh.

Considering that the best flat-plate system installed to date (at PVUSA) is only 10.7% efficient at the array level (maybe 12–13% at a module level), one can interpolate between the figures in Table 3.6 and estimate that the best modules today would have to be available for about $47/m^2$. It is a safe estimate that the PVUSA system cost more than 10 times that. From that perspective, there is a long way to go.

Although the goal of $.08/kWh is alluring, it should not be interpreted as an absolute requirement for the widespread use of large PV systems. EPRI and the PVUSA team have studied an application where 500–1000 kW PV plants could be used to reinforce a utility distribution system. For such an application, PV systems located near a load demand center could preclude or forestall the installation of larger distribution lines. In this scenario, the estimated COE could be $0.16/kWh and still be economically attractive. For this situation, the module cost could be $167/m^2$ ($1.11/W) for 15% modules.

Also, EPRI is sponsoring a program to identify other early utility applications for PV systems. They have located over 25 utilities that are already using PV power for some 60 different applications [18]. The intent of the program is to advertise the fact and to educate other utilities to the advantages of PV power. Thus, while central station systems may not be a factor for many years, PV is destined to be used more and more by utilities and the public. Many foresee a large number of small distributed systems that would reduce demand and be economical with only slight reductions in today's module cost.

3.2.9   *Will there be large PV systems in our future?*

Not soon. Unless there is a significant change in the price of fossil fuel or some dramatic political action that would levy a stiff tariff on the utilization of fossil fuel, large PV systems will remain a promise and not a reality. The cost reductions will not be easy to realize. PV companies have lost money for years. The increased demand for modules in the last 18 months (for small stand-alone systems) has given them a chance to increase prices. They are having trouble supplying enough modules; backlogs are common, and the average module selling prices have increased by about 10–20% in the US since 1989. Manufacturers are cautious about making the huge investments required to double or triple the manufacturing capacity without more interest from major customers like electric utilities — and this interest is not forthcoming. The most active utility in the US, Pacific Gas & Electric, may install 1 MW in the next few years. Hopefully, the Department of Defense will install some large (200 kW) hybrid PV/diesel systems but there are no firm plans yet. DOE is promoting the Solar 2000 strategy that calls for 1500 MW of installed systems by 2000 but this has no more credence than the 1981 plans for $0.70/W before 1986 [19]. The last thing manufacturers plan to do now is to flood the market and force module prices down.

Is there cause for alarm because few central station systems will be built this decade? No, they will be built if and when the economic conditions make them attractive. The technology is proven, but it is unreasonable to expect the cost of PV, based on today's dollars, to decrease to a point where large systems will be installed.

Other factors will drive the market. Many in the PV arena rebel at the fact that they cannot control their own destiny but gigawatts of PV will not be installed until the price of fossil fuel reflects a larger part of its cost to society. An awareness of the environmental consequences of our "way of life" in the last part of this century is growing among US citizens. There is a desire for others to do something about it, but there is scant indication that the average citizen is ready to do something about it himself. Until that occurs, we won't see central station PV systems in the US.

3.2.10   *References*

[1] L. N. Dumas and A. Shemka, "Field failure mechanisms for photovoltaic modules," *Proceedings of the 15th IEEE Photovoltaic Specialists Conference*, Kissimmee, FL, May 1981.

[2] S. E. Forman and m. p. themelis, "review of pv Module Performance at doe/mit lincoln Laboratory Test Sites During the Period 1977–1982,"

Proceedings of the 16th IEEE Photovoltaic Specialist Conference, San Diego, CA, September 27–30, 1982.

[3] Personal communication with Bill Priest, WBNO Radio, Bryan, Ohio, September 1991.

[4] K. L. Biringer, J. F. McDowell, C. B. Rogers, and D. E. Haskins, "Intermediate photovoltaic system/utility interface experience," *Proceedings of the 15th IEEE Photovoltaic Specialists Conference*, Kissimmee, FL, May 1981.

[5] V. V. Risser, "PV power system performance and reliability", *Proceedings of the 19th IEEE PV Specialists Conference*, New Orleans, LA, May 1987.

[6] D. E. Collier and T. E. Key, "electrical fault Protection for a Large photovoltaic power plant inverter," *Proceedings of the 20th IEEE PV Specialists Conference*, Las Vegas, Nevada, September 1988.

[7] D. D. Sumner, C. M. Whitaker, and L. E. Schlueter, "Carrisa plains photovoltaic power Plant 1984–1987 performance," *Proceedings of the 20th IEEE PV Specialists Conference*, Las Vegas, Nevada, September 1988.

[8] PVUS Project Team. 1989–1990 PVUSA Progress Report, 1991.

[9] M. L. Whipple, "1991 DOE/Sandia crystalline photovoltaic technology project review meeting," *Proceedings of the Annual Project Review Meeting*, Albuquerque, NM, July 1991.

[10] Personal communication with Greg Christianson, September 1991.

[11] R. Rosenthal and V. Risser, *Photovoltaic System Performance Assessment for 1988*, EPRI GS-6696, January 1990.

[12] V. Risser and K. Stokes, *Photovoltaic Field Test Performance Assessment for 1987*, EPRI GS-6251, March 1989.

[13] V. Risser and K. Stokes *Photovoltaic Field Test Performance Assessment for 1986*, EPRI AP-5762, March 1988.

[14] V. Risser and K. Stokes, *1985 Performance Evaluation of the Hesperia Photovoltaic Power Plant*, Electric Power Research Institute, March 1987.

[15] S. J. Durand, D. R. Bowling, and V. V. Risser, "Lessons learned from testing utility connected PV systems," *Proceedings of the 21st IEEE PV Specialists Conference*, Orlando, FL, May 1990.

[16] C. Jennings, "Outdoor versus rated photovoltaic module performance," *Proceedings of the 19th IEEE PV Specialists Conference*, New Orleans, LA, May 1987.

[17] E. A. DeMeo, "Photovoltaic for bulk power applications: cost/performance targets and technology prospects," *Proceedings of the 10th European PV Solar Energy Conference*, Lisbon, Portugal, April 1991.

[18] J. E. Bigger, *Cost-Effective Photovoltaic Applications for Electric Utilities*, U.S. Photovoltaic Activities for CRIEPI and Japanese Utility Representatives, Electric Power Research Institute, May 1991.

[19] P. D. Maycock, "Overview of the US photovoltaic program by an Ex-DOE person," *Proceedings for 15th IEEE Photovoltaic Specialists Conference*, Kissimmee, FL, May 1981.

3.3   Discussion following Vern Risser's presentation

Question: What do you think of the prospects for very high concentration cells?

Risser: Alpha Solarco makes a 400× point focus system, but it is too early to say how successful it is. My guess, however, would be that high concentration stands a better chance than thin film technology for getting us down to the $1/Wp level.

Question: Are you thinking in terms of silicon or gallium arsenide cells?

Risser: Silicon. I don't believe that GaAs will be developed for systems of the 100MW scale.

Question: Why are you so negative about thin film technology?

Risser: I am only critical of the performance of those thin film modules that have been demonstrated to date.

Question: How do you explain the poor field efficiency of amorphous silicon modules?

Risser: High-efficiency a-Si cells have usually been achieved in the laboratory by carefully producing a relatively small-area cell and, in some cases, selectively cutting from this an even smaller cell corresponding to the area of highest efficiency. From here, the scale-up process to industrial mass production is a far from trivial problem.

Faiman: Is module browning only caused by mirrors?

Risser: The phenomenon is not fully understood. At SWRES, we have observed browning on modules that have been exposed to natural sunlight for 10 years and on newer modules that have been exposed to mirror-enhanced sunshine. It may simply be that the mirrors produce accelerated aging on the ethylene vinyl acetate (EVA) encapsulant. Incidentally, the coloration on your Paz system here is what we refer to as "shadowing" rather than "browning". Shadowing tends to decrease the efficiency of modules by at most a few percent. By way of contrast, the browning produced on the mirror-enhanced modules at Carrisa Plain resulted in a considerable loss in efficiency.

Spiewak: Why are there no Japanese photovoltaics manufacturers in the PV America program? Are their modules inferior?

Risser: Their modules appear to be of similar quality to those of US or European manufacture. I think it is probably due to a policy of the funding agency to promote American industry. In fact it is possible that one Japanese manufacturer, which claims to produce more than 50% of its output in the USA, may qualify for inclusion.

Arbib: When you talk about home distributed systems do you mean the ones which will sell electricity to the grid?

Risser: In principle, yes, but for the near term I do not believe that it would be very profitable for the homeowner to sell PV power to the grid. An economically sized home system would probably provide 50—60% of the owner's requirements with very little surplus for sale purposes.

Arbib: Is there a danger that privately owned systems of this kind might cause disturbances to the grid?

Risser: If you mean disturbances to the smoothness of the waveform on the grid, I believe that such problems are easy to overcome. In fact, the degree of matching that has already been achieved is impressive. In a so- called "islanding" exercise I witnessed in Japan, 200 homes were suddenly disconnected from the grid and their respective PV systems were made to take over. I was in one of the houses when this happened and did not notice a flicker in any of the lamps or appliances.

A possibly more serious "disturbance" might arise from the fact that when a line comes down, the utility needs to be sure that the fallen cables are not live before their servicemen can go to work. Active PV systems at the other end may present some logistics problems, but I am sure that these would be easy to overcome.

Ultimately, there will also be an institutional problem to face. If PV systems were ever to penetrate the market to an extent that utilities would need to maintain spinning reserves purely for cloudy days, then an economically absurd situation would have arisen. But there is a long way to go before such problems need to be faced.

Roy: Pumped storage could help overcome that problem.

Risser: Possibly. An appropriate kind of storage mechanism would need to be available at the location of each power station.

Gordon: You did not mention developments in the US space program.

Risser: Your observation is correct. I have deliberately restricted my review to terrestrial systems, where cost-effectiveness is the overriding consideration.

Arbib: What are the current price of PV modules and the balance of system costs?

Risser: Module costs have actually risen slightly during the last 2 years. They are currently around $6.50/Wp.

EPRI/s current estimates for BOS costs are $50/m^2$ for flat plate and $100/m^2$ for concentrator systems. (If one takes 10% as the nominal system efficiency, then $50/m^2$ works out at $0.50/Wp.) These estimates are probably somewhat optimistic. I believe the cost would be this low only if a large system were installed using a well thought out plan and some automated procedures.

Chapter 4

1993

4.1 Editor's Foreword

Mark Koltun's keynote presentation at the *Fifth Sede Boqer Symposium on Solar Electric Power Production* nicely complemented the earlier reviews of Fred Treble on PV in Europe and Vern Risser on PV in the USA. However, this presentation was more than a review of PV developments in the former USSR. The reason is that in contrast to the situation in the West, PV research in the former Soviet Union was highly secret and what little was published was done so in the Russian language. Consequently, although some of the Soviet work had been translated, piecemeal, into English, Koltun's review was essentially the first opportunity for Western listeners to receive a comprehensive picture of the manner in which Soviet scientists developed PV cells for their space program, and the lessons they learnt that are of relevance to the terrestrial use of this technology for power production purposes.

Tragically, Mark Koltun drowned while swimming in the Mediterranean Sea during one of his visits to Israel. The present text of his Sede Boqer keynote lecture is a slightly polished version of the original, prepared by this editor for a special issue of the journal *Solar Energy Materials and Solar Cells* (Vol. 44, 1996, pp. 293–317) that was dedicated to his scientific achievements. The editor is indebted to Elsevier Press for permission to reproduce the polished version.

4.2   History of solar cell development in the Soviet space program and the terrestrial potential for this technology (Professor Mark M. Koltun)

A keynote lecture presented by Mark M. Koltun (Photovoltaic Laboratory, Renewable Energy Sources Department, Krzhizhanovsky Power Engineering Institute, 19 Leninsky Prospect, Moscow 117927, Russia, and Solar Energy Research and Educational Department, Institute of New Technologies, 10 Nyzhnaya Radishchevskaia Street, Moscow 109004, Russia)

4.2.1   *Introduction*

Solar cells have been found to be reliable and efficient as sources of electric power for space vehicles, also as sources of information on the spectral composition and direction of solar rays, and as simple elements for the automatic orientation of systems in space vehicles.

Solar cells can be widely used not only in the relatively small independent systems used in space, but also for domestic and industrial purposes. This is clear from the development and successful testing of the first solar houses, and the use of photovoltaic sources and power supplies for radio receivers, television sets, radio stations, signaling equipment, and buoys. Solar electric power stations employing thermal as well as photovoltaic converters are interesting and very promising.

Solar cells based on the photoelectric effect in semiconducting structures with a barrier layer (photovoltaic effect) can directly convert the optical radiation incident upon them into electric power. They are thus electric power generators (in contrast to photoresistors and photocells exploiting the external photoelectric effect) and do not require an external voltage source.

Since the discovery, in the middle of the last century, of the photoelectric properties of selenium, and the development at the turn of the century of the first photoelectric converters of radiation into small electric signals (based on selenium and copper/copper oxide heterosystems), there have been frequent attempts aimed at increasing the efficiency of such converters and exploiting them as sources of useful electric increasing power. Improved technology, and also the use of optical filters, resulted in selenium photocells whose spectral sensitivity was practically the same as that of the human eye. Improved selenium photocells found extensive applications as exposure meters in photography and cinematography. However, the efficiency of such photocells did not exceed 0.5%. The successful application of the photoelectric method of energy conversion began only after the advent of the band theory of the electronic structure of semiconductors, the development of methods for their purification and controlled doping, and the elucidation of the dominant role of the barrier layer on the separation boundary between semiconductors with different types of conductivity. A brief report on the development of the silicon solar cell with an efficiency of about 6% appeared in 1954, and by 1958, both Soviet and American satellites carried silicon solar batteries to supply electric power for the electronics. Since then, the efficiency of solar cells has been substantially increased. This has been facilitated by better understanding of the physical phenomena occurring in solar cells, by the introduction of continuously improving fabrication techniques, and by the development of new and improved cell designs relying on a variety of semiconductor materials.

These studies were, in turn, based on the systematic theory of photoelectric phenomena in semiconductors, established in the 1930s and 1940s. In the USSR, this branch of semiconductor physics has developed as a result of the pioneering work of academician A.F. Ioffe and his school, who extended our understanding of the nature of photoconductivity and photoelectric phenomena in semiconductors and semiconducting p–n junctions [1].

Work done in the Soviet Union and elsewhere on the development of photoelectric devices converting radiant energy, including solar energy, into electric power has been systematically reviewed in scientific monographs [2–5].

The efficiency of solar cells depends on the optical and electrophysical properties of the semiconducting material (reflection from the surface, photoionization quantum yield, diffusion length of minority carriers, and position of the fundamental absorption band), as well as on the characteristics of the p–n junction (origin of reverse current, height of potential barrier, and width of the space charge region), the so-called geometric factor (ratio of carrier diffusion length and depth of the p–n junction), and the dopant concentration on either side of the p–n junction. It is clear that the shape of the current–voltage characteristic and the output power depend on the series resistance, which, in turn, depends on the resistance, thickness, and dopant concentration in the semiconductor, as well as on the shape and position of the current contacts. The desire to reconcile these frequently conflicting requirements, and to find the optimum technological compromise, has led to the preferred use of the planar design for solar cells (see Fig. 4.1(a)).

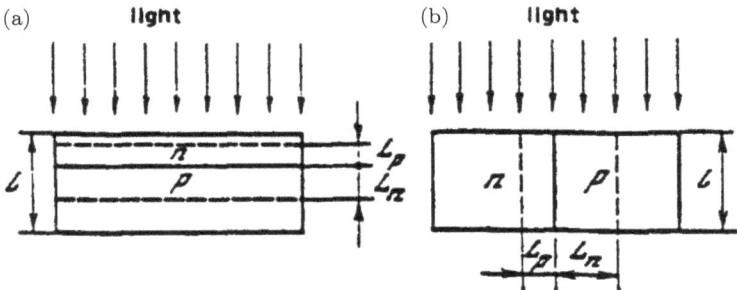

Figure 4.1:   Diagram illustrating the position of the p–n junction in a semi-conducting crystal under (a) perpendicular and (b) parallel incidence of optical radiation. L_n and L_p are the diffusion lengths of minority carriers in the p- and n-type regions, respectively, and l is the depth of penetration of light into the semiconductor.

With small modifications (introduction of drift fields and of an isotype barrier on the back contact; replacement of a solid back contact with a grid; texturing of the surface of the semiconductor and its coating; and deposition of a reflecting layer on its rear surface), this design has been with us without any substantial change for many years, at any rate in the case of solar cells made from monocrystalline silicon with a homogeneous p–n junction, and continues to dominate space and terrestrial applications.

The upper surface of the silicon cell, which faces the incident optical radiation, is very thin and highly doped (impurity atom concentrations of up to 10^{20}–10^{21} cm^{-3}), for example, with phosphorus atoms, so that it becomes an n-type. The p-type region of the semiconductor is usually lightly doped, for example, with boron (usually while the crystal is being grown), up to an impurity atom concentration of 10^{16}–10^{17} cm^{-3}. The outer surface of the solar cell is covered by a grid of current-taking strips occupying 5–7% of the total area, while a solid or grid contact is placed at the rear of the cell.

The minority carriers separated by the p–n junction field must enter the external circuit (load). In the upper, n-type, region of the semiconductor, which faces the incident light, the minority carriers move along the layer, whereas in the p-type base region (see Fig. 4.1(a)), they move across the layer. The diffusion length of the minority carriers in the highly doped n-type upper layer is usually 0.2–0.6 μm, while in the base layer, it is 100–200 μm. These figures depend on the impurity concentration and the thermal treatment (the number of cycles, rate of heating and cooling, and maximum temperature) applied to the crystal while it is being grown, and to the solar cell during the fabrication process (for example, during the thermal diffusion of dopants and the deposition and strengthening of the antireflective coatings).

It is important to note that the numerous thermal treatments applied to the semiconducting layers at different technological stages of solar cell fabrication necessarily involve the entry of undesirable impurities and recombination centers which affect the optical and electrophysical parameters of semiconducting material. This means that the parameters of the semiconductor are best determined at the end of the technological cycle. This is usually done by calculation from the output characteristics of solar cells, for example, the current–voltage characteristic or the spectral sensitivity, or from certain other more specific curves, for example, the capacitance–voltage and current–irradiance characteristics (capacitance as a function of applied voltage, and basic photoelectric parameters as a function of irradiance).

The diffusion length in the doped layer is small, so that we have to use a shallow p–n junction (0.3–0.6 μm in modern mass-produced solar cells). To ensure that all the incident solar photons with $h\nu \geq E_g$ are absorbed, the thickness of the base region must be not less than 200 μm. The resistance of the base region is low: current flows across a layer with a relatively large cross-section, toward the solid or grid base contact fused into the silicon at $750 - 800°C$ in an inert atmosphere. The first layer of the contact is often made of aluminum, which is a p-type impurity, in order to reduce the metal–silicon junction resistance (p-type). Aluminum is deposited by evaporation in a high vacuum, or in the form of aluminum-containing paste with an organic binder. The aluminum layer is then covered by a film of titanium, palladium, or silver (nickel is an alternative choice), and a layer of tin or lead solder.

The high layer resistance of the n-type top silicon layer, which, as a rule, is in the range 50–100 $\Omega/\square$, is effectively reduced by placing on the outer surface a dense metalized grid of current contacts, made from the same material as the back contact (with the exception of the aluminum layer, which is unnecessary in the case of a contact grid with the n-type layer). Another problem that is encountered during the fabrication of the top current contact is that it is essential to produce a satisfactory (non-rectifying) contact that will not pierce the very thin doped layer during deposition and subsequent treatment. Experiment shows that the deposition of a metal layer over the entire outer surface, followed by the formation of the contact figure by etching, gives rise to the appearance of shorting microregions, which reduce R_{sh}, and to an increase in reverse current I_o in both single-crystal and thin-film solar cells. This can be avoided by depositing the contact strips through a metal mask, or through windows in a layer of polymeric photoresist or antireflective coatings, or directly through the antireflective coating. At any rate, it is essential to ensure that the metal and the doped layer come together only at the intended point of contact.

For a layer resistance of between 50 and 100 $\Omega/\square$ on the outer surface of a 2×2 cm silicon solar cell, it is sufficient to produce one contact in the form of a strip, 0.5–1.0 mm wide, on the side of the cell, with between 6 and 12 outgoing current-collecting contact strips, 0.05–0.1 mm wide, attached to the main strip. This reduces the contribution of the doped layer to the total series resistance R_s of the cell down to 0.15–0.2 Ω. However, for very shallow p–n junctions ($l = 0.15 - 0.2\,\mu$m), the layer resistance rises to 500 Ω/□, and the number of contact strips on a 2×2 cm solar cell must be increased to 60. The necessary low resistance of a 15–20 μm contact strip is then achieved by subsequent electrochemical deposition of a silver layer of thickness up to 3–5 μm.

The solar cell with a p–n junction in a homogeneous semiconductor is made from a homogeneous semiconductor whose basic optical and electric properties (including the band gap) are the same at all points within the volume. Structures and solar cells based upon them are referred to as graded-gap systems if the width of the band gap varies, e.g., decreases with distance into the crystal, owing to the continuous variation in the chemical composition of the material, and the p–n junction lies at a certain depth. The junction may be located on the boundary between the semiconducting layers of different band gaps (it is usually called a heterojunction), or it may be in one of the layers, usually in the lower layer with the smallest band gap. The upper layer of the wide-gap material is then merely an optical window that transmits the incident light onto the p–n junction. On the other hand, the boundary between the wide-gap and narrow-gap materials with similar lattice constants (as in the case of GaAlAs-GaAs and Cu_2S–CdZnS) has a low rate of carrier recombination. Since, in solar cells with a p–n junction in a heterostructure, recombination on the upper boundary turns out to be sharply reduced, the carrier collection efficiency (especially in the short-wave part of the spectrum) is higher, and the efficiency of such cells can be very high.

Many problems connected with the efficient solar cells' technology and design (for cells made from different semiconducting materials) were solved in the USSR, and many original inventions were made, but only three of them became known in the West. Here are three abstracts from the excellent book by Hans Rauschenbach describing these works [2]:

"In the US, the original material used for solar cells was n-type silicon, while in the USSR it was p-type material. Silicon of the p-type was used by the Russians (1956) for two reasons: (1) to scientifically

contrast the US work; and (2) because p-type material was cheaper in
the USSR than was n-type. It was later found that cells made from
p-type silicon were more resistant to corpuscular radiation as found in
space than were cells made from n-type material. Thus, after discovery
of the Van Allen radiation belts, US solar cell production after I960
switched over to diffusion of n-layers into p-type silicon." "Photovoltaic
installations in the USSR date back to the 1960's. Attractive especially
for use in the remote, semi-arid southeastern areas of that country,
a variety of systems, ranging in size from 1 to 500 watts output for
powering irrigation pumps, water gates, communications equipment,
and light buoys on water-ways, have been installed. Some installations
utilize solar concentration up to 1000 times, aided by highly accurate
tracking equipment. Array configurations include fixed and sun-tracking
flat plate arrays with and without low-concentration ratio mirrors, and
long, narrow arrays contained in transparent, sealed glass and non-glass
tubes. The tubes are dried out and filled with an inert gas before sealing.
The solar cells are made from silicon, cadmium sulfide, and cadmium
telluride. A typical temperature rise of $30°$ C between the solar cells
and the ambient air temperature has been observed." "After Russian
workers reported in 1971 on an GaAlAs/GaAs cell structure, worldwide activities sprang up in several research centers. A thin GaAlAs
layer, known as the "window", is deposited on a p-on-n GaAs cell and
significantly reduces surface recombination losses. The new cell design
paved the way for increased cell efficiency, exceeding 16% in 1977 at AM0
in 2×2 cm sizes. Efficiencies in excess of 23% under AM1 illumination
for 0.25 cm^2 cells was claimed."

I will further try to add more details to these important but
too brief remarks of Hans Rauschenbach. I will not tell about the
everyday work done in every advanced photovoltaic country (and
also in the USSR), but will concentrate on several of the ordinary
investigations made in Russia and the former USSR.

4.2.2   *High-efficiency silicon solar cells with a drift field in the doped region*

Modern solar cells have internal electric fields that are random
in character and are a consequence of the fabrication technology
employed. The problem that had to be faced was, therefore, to
find the impurity distributions that would substantially enhance the
efficiency of the collection of carriers from the doped layer and, at
the same time, could be produced by well-established technology.

The idea was to consider the possibility of a solar cell in which the doped layer consists of two regions with a different impurity concentration (Fig. 4.2). A jump in the potential, $U_E = (kT/q)\ln(N_1/N_2)$, occurs across the separation boundary between the two regions and the concentrations obey the inequality $N_1 > N_2$, so that the electric field across the separation boundary between regions I and II points toward the p–n junction. The first step will be to optimize the parameters of the doped layer with respect to the photocurrent and power by taking into account the series resistance.

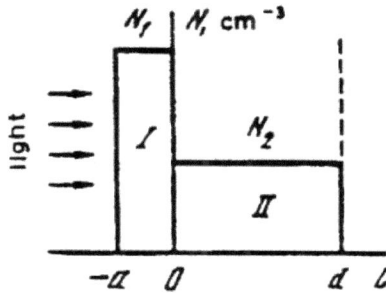

Figure 4.2: Two-layer model of the doped region in a solar cell. The dashed line represents the p–n junction at a depth $l_d = a + d$.

The mobility and the carrier diffusion length as functions of impurity concentration can be taken in the form

$$\mu_n, \mu_p \sim N^{-\gamma}, L_p \sim N^{-\beta}$$

which agrees with experimental data for $\beta = \gamma = 1/2$ to a sufficient degree of precision.

We have calculated the carrier collection from doped layers where L_{P1} and L_{P2} are the diffusion lengths of minority carriers in regions I and II, respectively, and a and d are the widths of these regions (see Fig. 4.2). We also derived expressions for the photocurrent and output power of solar cells with an electric field in the doped layer.

The photocurrent from the doped layer was obtained as a function of the thickness of the region with enhanced carrier concentration for different values of a and d (Fig. 4.2). The dopant concentration in the p–n junction was $N_2 = 10^{17} - 10^{18}$ cm^{-3}. The carrier concentration

N_1 on the surface was assigned a number of values in the range 10^{18}–10^{21} cm^{-3}, where the maximum value $N_1 = 10^{21}$ cm^{-3} corresponded to the solubility limit of phosphorus in silicon.

Our calculations showed that the maximum photocurrent from the doped layer was obtained within the above limits for $a = 0.05\mu$ m and $N_1/N_2 = 100$.

However, subsequent analysis revealed that, for fixed N_1 and N_2, the useful power can be a maximum for $a < 0.05\,\mu$m. The point was that, to produce a high photo-emf in real solar cells, the necessary carrier concentration N_2 in the p–n junction had to be $10^{17} - 10^{18}$ cm^{-3}. For such concentrations, the spreading resistance of the thin $(0.1\,\mu\text{m})$ doped layer was found to be relatively high, but could be reduced (for the same shape of contact on the working surface) by expanding region I with a higher impurity concentration.

Figure 4.3 shows the power produced by a solar cell with antireflective coating and optimized doped-layer parameters described by

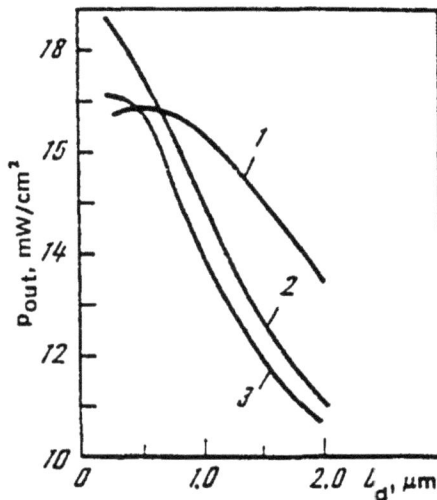

Figure 4.3: Calculated output power as a function of the depth l_d of the p–n junction in a solar cell in which the electric field is: (1) power optimized for $N_1 = 10^{19}, N_2 = 10^{17}$ cm^{-3}; (2) power optimized for $N_1 = 10^{20}, N_2 = 10^{18}$ cm^{-3}; (3) uniform.

the above model. The figure also shows the results for an ordinary solar cell, in which the doped layer has a uniform electric field (the concentration falls from $5 \times 10^{20}\,\mathrm{cm}^{-3}$ on the surface to $10^{17}\,\mathrm{cm}^{-3}$ at the p–n junction, the impurity being an exponential function of depth).

When the impurity concentration in the doped layer is $N_1 = 10^{19}$ and $N_2 = 10^{17}\,\mathrm{cm}^{-3}$ (curve 1), the power produced by the solar cell based on the above model exceeds the power delivered by an ordinary cell for $l_d < 0.6\,\mu\mathrm{m}$. This layer is particularly convenient in the case of a deep p–n junction. Actually, when $l_d = 0.7\,\mu\mathrm{m}$, the increase in power amounts to 5% and the corresponding figures for $l_d = 1.0, 1.5,$ and $2.0\,\mu\mathrm{m}$ are 17%, 27%, and 28%, respectively. Doped layers with composite impurity distributions can be used to achieve higher useful power for greater depths of the p–n junction as compared with the exponential impurity distribution. For example, $P = 16\ \mathrm{mW/cm^2}$ corresponds to $l \cong 0.7\,\mu\mathrm{m}$ (curve 3) and $l_d \cong 1.2\,\mu\mathrm{m}$ (curve 1).

When the impurity concentration in the doped layer is raised to $N_1 = 10^{20}, N_2 = 10^{18}\,\mathrm{cm}^{-3}$ (curve 2), the increase in power as compared with the case of the uniform field is 4–7% for all values of l_d. A somewhat greater increase in power (up to 10%) is observed for $l_d < 0.5\,\mu\mathrm{m}$. Thus, if solar cells can be made with reliable contacts for p–n junction depths less than $0.5\,\mu\mathrm{m}$, better results can be achieved by producing the doped layer with a stepped distribution of high concentration (to reduce the series resistance).

The above theoretical results were examined experimentally using a doped layer in which the stepped impurity distribution was established by thermodiffusion technology, which is widely used in the fabrication of silicon solar cells. Diffusion was produced by the box method.

Calculations and experiments have shown that the porous oxide film initially produced on the surface of silicon by anode oxidation can be exploited even in the case of a single diffusion process in order to produce the two-layer doped region [5]. Part of the diffusant, for example, phosphorus, passes through the pores and forms a region of low impurity concentration in the p–n junction. The impurity current retained by the oxide layer produces a thin layer with enhanced impurity concentration on the surface. By varying the porosity of the film and controlling the diffusion time and temperature, it is possible to control quite smoothly and accurately the impurity distribution in the doped region.

A p–n junction in which the depth of the doped layer is 0.9–1.3 μm can be produced by optimized single thermodiffusion through an oxide film of the necessary porosity, which is first deposited on the system. The impurity distribution can thus be made to take the form of the two regions of high and low concentration (curve 1, Fig. 4.4).

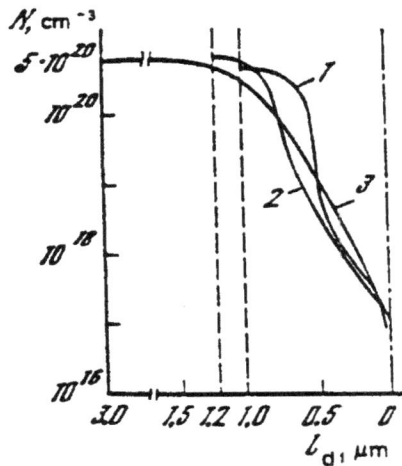

Figure 4.4: Measured concentration of phosphorus in silicon as a function of depth between the surface (dashed lines) and the p–n junction (dot–dash line). The layer is produced by thermodiffusion under different conditions: (1) single thermodiffusion through an impeding oxide layer of a particular porosity ($l_d = 1.0\,\mu$m); (2) double thermodiffusion through an oxide layer on the surface ($l_d = 1.2\,\mu$m); (3) thermodiffusion without preliminary oxidation of the surface ($l_d = 1.2\,\mu$m after chemical etching of the doped layer).

Another possibility is double doping. This was achieved by using selected silicon discs with a 3-μm doped layer, produced by thermodiffusion, in which the impurity distribution was as shown by curve 3 in Fig. 4.4.

The diffused layer was etched out to a depth of 0.5–0.6 μm and this was followed by secondary doping by single thermodiffusion.

The resulting p–n junctions were at a depth of 1.0–1.2 μm from the surface, and it was found that the impurity concentration changed by two orders of magnitude over the depth range 0.3–0.7 μm (curve 2, Fig. 4.4). The impurity concentration profile was determined from conductivity measurements using the four-probe method, and layer-by-layer anode etching. The depth of the p–n junction was determined by the grooving method.

Current contacts were deposited on the silicon wafers in the usual way, and the characteristics of the resulting photocells were investigated.

The experimental solar cells have enhanced spectral sensitivity in the short-wave part of the spectrum (curves 1 and 2, Fig. 4.5), which depends on the efficiency of collection of carriers from the doped region.

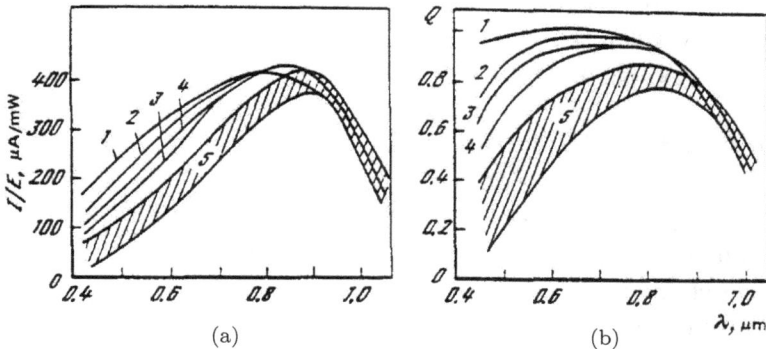

Figure 4.5: Spectral sensitivity (a) and collection coefficient (b) for uncoated solar cells fabricated by different methods: (1–3) as in Figs. 4.3 and 4.4; (4) shallow thermodiffusion in a gas flow (exponential distribution of impurities; $l_d = 0.6 \mu$m); (5) deep single thermodiffusion (without chemical etching after diffusion; $l_d = 3.0 \mu$m).

For example, for $\lambda = 0.5 \mu$m, the experimental solar cells with stepped impurity distribution in the doped layer have $I/E = 220$–250μA/mW, whereas for cells with doped-layer thickness of the order of 3 and 1.2 μm (the corresponding impurity distributions are described by the horizontal and sloping parts of curve 3 in Fig. 4.4), the spectral sensitivity lies in the range 50–125 μA/mW (region 5, Fig. 4.5) and 170–180 μA/mW (curve 3, Fig. 4.5). Even for elements with a very shallow p–n junction (0.6 μm) and exponential impurity distribution (curve 4, Fig. 4.5), the sensitivity at $\lambda = 0.5\,\mu$m does not exceed 200 μA/mW. Solar cells with p–n junction depth $l_d = 0.6\,\mu$m (curve 4), produced by low-temperature diffusion, have a near-exponential impurity distribution in the doped layer, with a concentration drop from 5×10^{20} on the surface to 10^{16} cm^{-3} at the p–n junction. Comparison of curves 1–4 will show that the enhanced sensitivity of the experimental solar cells in the short-wave part of the spectrum (curves 1 and 2) can be explained by the dominating effect (as compared with the deterioration in the minority-carrier diffusion parameters in the region of enhanced concentration) of the built-in drift field of complex configuration.

The current–voltage characteristics of solar cells with the two-layer structure of the doped region are also much better than those of ordinary cells. The load current drawn per unit useful area of the solar cell when the p–n junction is at a depth of 1.0–1.2 μm is greater by 9–17% when compared with a cell using an exponential impurity distribution in the doped layer, and this can be regarded as a sufficient confirmation of the calculated results (see Fig. 4.3).

The proposed stepped distribution of the impurities is thus seen to result in a considerable improvement in the current–voltage and spectral characteristics of solar cells, even when the p–n junction is relatively deep ($l_d = 1.2\,\mu$m), so that one can not only increase the efficiency of the cells but also use simple, cheap, and reliable contacts produced by chemical deposition of nickel. The problem of producing reliable ohmic contacts and of automating their deposition (as well as making them cheaper) is one of the most complex problems in the modern technology of solar cell fabrication.

4.2.3 *Silicon solar cells with a drift field in the base and an isotype barrier on the back contact*

When the base layer of a solar cell, for example, a p–type layer, is doped non-uniformly and the acceptor concentration in the p–n junction is lower than that within the body of the layer, an electric field appears [6] and assists in the collection of excess carriers produced by light in the base layer (both the diffusion and drift collection mechanisms are then found to be operative). It has frequently been noted that an impurity gradient is a precondition for a drift field. On the one hand, this reduces the open-circuit voltage due to the increase in the reverse saturation current when the potential barrier is reduced (by reducing the dopant concentration in the base of the p–n junction) and, on the other hand, this leads to a considerable deterioration in the diffusion length and lifetime of minority carriers (when the dopant concentration is increased in base layers that are distant from the p–n junction). Both phenomena work against the improvement in the collection coefficient due to the drift field in the base layer (which is usually uniformly doped) because of the associated inhomogeneous doping. For a relatively small change in the concentration across the base layer (10^{17} cm^{-3} in the p–n junction and $10^{18} - 10^{19}$ cm^{-3} within the base), it is possible to produce an increase in the efficiency and long-wave spectral sensitivity of silicon cells by introducing a drift field while maintaining at a reasonable level the diode parameters of the p–n junction and the lifetime of minority carriers in the base [7].

The experimental realization of this model by slow diffusion of an impurity into the base wafer of a solar cell was found to be laborious and time-consuming. The technology of exodiffusion of impurities in a vacuum from a base wafer doped in advance, which we have developed, was equally complex. The use of rapidly diffusing lithium found practical application in the fabrication of space cells with radiation-resistant properties, not only because the drift field of a sufficient intensity could be produced in the base layer by a relatively simple technology, but also because lithium neutralized the recombination centers produced by radiation. Cheap terrestrial cells can be fabricated by growing an epitaxial silicon layer with a graded impurity concentration on a single-crystal silicon wafer and then introducing the p–n junction (by thermal diffusion or by deposition of an epitaxial highly doped film with a different type of conductivity) from the side of the epitaxial layer. The cost reduction is achieved because the epitaxial film is deposited on metallurgical silicon, which is 100 times cheaper than semiconductor-type silicon. The automatic doping of the epitaxial layer, while it is being grown by impurities from the substrate, produces the necessary impurity concentration gradient and, hence, the drift field.

Solar cells with a strong drift field in the base were soon replaced by cells with a sharp isotype p–p$^+$ or n–n$^+$ junction on the metallic back contact, similar to the two-layer model considered in the previous section of this paper.

Near-intrinsic silicon could be used to produce high-efficiency solar cells by diffusing n- or p-type impurities from either side of the silicon wafer so as to produce the p–n junction at the necessary distance from the surface and, at the same time, achieve the optimum impurity gradient on the other side of the wafer. Experience gained in the fabrication of n$^+$-p–p$^+$ or p$^+$-n–n$^+$ structures [8] shows that it is much simpler to produce a very thin isotype p–p$^+$ or n–n$^+$ junction on the metallic back contact than to establish a strong drift field, and that this is virtually as good from the point of view of minority-carrier collection from the base layer. The potential barrier on the isotype junction, produced by doping the base from the back, reflects the minority carriers from the back contact, increases their effective diffusion length, and reduces to practically zero the rate of surface recombination on the separation boundary between the base and the metallic contact. There is also some reduction in the reverse saturation current of the cells. The doped back layer is produced by thermodiffusion, ion bombardment, or the implantation of aluminum (in the case of the p-type layer), followed by thermal treatment. The depth of the doped layer is usually between 0.2 and 0.5 μm, and the impurity distribution is practically the same as in the upper doped layer of solar cells.

The advantages of solar cells with isotype junction on the back surface become significant when the diffusion length of minority carriers on the base is greater than the thickness of the base layer or at least equal to it. This requirement means that the base layer must be a sufficiently pure semiconducting material of high enough resistivity, or the thickness of the base must be reduced to a value that is less than the diffusion length of minority carriers in the material.

Solar cells with the p–i–n or p^+-i-n^+ structure and their modifications have exceptionally high sensitivity at long wavelengths. The current–voltage characteristic of such cells is nearly square because the high level of illumination under exposure to solar light ensures that the voltage drop across the base region is reduced to a minimum (the non-equilibrium carrier concentration produced under illumination of the high resistivity base is much greater than the concentration of equilibrium carriers). The large initial diffusion length of minority carriers in the high-resistivity material extends the life of such cells in the Earth's radiation belts [9].

4.2.4 *Solar cells transparent in the long-wave region beyond the fundamental absorption edge*

The basic possibility of such solar cells is assured by the transparency of any pure high-resistivity semiconducting material beyond the fundamental absorption band. However, if the base of the solar cell is made from a relatively pure material with a low dopant concentration, the upper layer must be doped to a concentration that is practically equal to the solubility limit for the donor or acceptor impurity in the given semiconductor, in order to reduce the spreading resistance of carriers separated by the p–n junction. Long-wave radiation will, of course, be strongly absorbed and reflected by this type of highly doped layer.

The low lifetime and diffusion length of minority carriers in the doped layer means that the thickness of the layer must be reduced to a value in the range 0.15–0.2 μm. Absorption of the infrared component of solar radiation ($\lambda = 1.1 - 2.5\,\mu$m) by a cell with a doped layer of this thickness does not exceed 1–3%. The tendency to reduce the depth of the p–n junction in modern solar cells has thus removed one of the obstacles to producing cells that are transparent in the long-wave region of the spectrum.

Two other obstacles, namely, absorption by the solid back contact and high reflection by the back surface of the cell, were overcome by replacing the solid back contact with a grid and depositing an antireflective coating with an optical thickness of 0.3–0.4 μm.

Calculations show that with a gridded back contact of a particular configuration, it is possible to preserve the series resistance and curve factor of the current–voltage characteristic of the transparent silicon solar cell practically at the level achieved for the conventional cell with a solid back contact. Similar results have been obtained for solar cells made from gallium arsenide.

Transparent solar cells made from silicon and gallium arsenide were used in the first practical version of the cascade cell [10]. The equilibrium working temperature of transparent silicon solar cells used in space is much lower than that of conventional cells, so that the integrated absorption coefficient for solar radiation decreases (as has been shown by direct measurement in space) to the range 0.72–0.73 rather than 0.92–0.93 (these are values characteristic for conventional cells with a solid back contact in the form of a fully reflecting metal) [11].

Silicon and gallium arsenide solar cells with a gridded back contact, which are transparent in the infrared beginning with $\lambda = 1.1\,\mu$m, were produced in the USSR [10, 11], whereas cells using Cu_2S–CdS thin-film structures were made in France [12]. Calculations have shown that the temperature of these cells in a geostationary orbit in space should be lower by $10 - 12°$ C and the output power should be higher by 5–6%.

Non-photoactive long-wave infrared radiation will not only be transmitted by the transparent solar cells, but also be reflected by its back surface. This can be done by depositing a highly reflecting film of, say, aluminum, copper, or silver on the back surface of the transparent solar cell free of the resistive current contact.

The reflector, in the form of the usual titanium–palladium–silver three-layer structure, can be deposited by evaporation in a high vacuum directly on the silicon surface, free from contact strips, or it can be produced simultaneously with the aluminum contact. However, the infrared reflection coefficient of this type of layer is reduced when the resistive aluminum contact is baked at high temperature. Figure 4.6 shows the spectral reflection coefficient of some of the high-efficiency solar cells with an aluminum back contact.

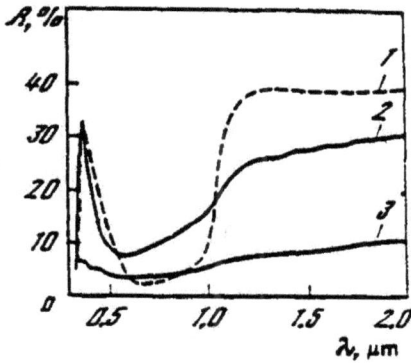

Figure 4.6: Reflection coefficient of n-p-type silicon solar cells with an aluminum back contact and an antireflective coating of tantalum pentoxide deposited on the outer surface of the cell after the following treatments: (1, 2) polished surface; (3) non-reflecting back surface produced by selective etching.

Two of the cells (1 and 2) had polished back and front surfaces, whereas cell 3 had a non-reflective back surface produced by spaced pyramidal etch pits. The front surface of all three cells was covered with films of tantalum pentoxide of different thicknesses. The baked-on reflecting aluminum contact could be used to increase the reflection by the solar cells, in the long-wave region beyond the fundamental absorption edge for $\lambda = 1.1$–$2.5\,\mu$m, to only 40%.

When the back contact is the three-layer titanium–palladium–silver deposit, the reflection coefficient in this region is no more than 20–30%, but can be increased to some extent by reducing the thickness of the titanium film. Solar cells with a nonreflecting back surface absorb practically completely not only radiation between 0.4 and 1.1 μm (the region of spectral sensitivity of the cell), but also the infrared radiation beyond the fundamental absorption edge. Transparent cells based on this principle cannot be produced.

It is much more convenient to increase reflection in non-photoactive parts of the spectrum by using a highly reflecting metal deposited on the silicon surface in the opening of the grid contact on the back. The silicon surface can be subjected to relatively slight heating (up to $150 - 200°$ C) to improve the adhesion of the layers while preserving at a high enough level the infrared reflection by the silicon–metal boundary. The spectral reflection of silicon solar cells with a three-layer coating (antireflective film consisting of zinc sulfide + organosilicon adhesive+protective glass) and different reflective layers (copper, aluminum, silver, nickel, and titanium) on the back surface without contact strips is up to 75–95%, despite the presence of selective absorption bands of the organosilicon compound between 1.1 and 1.5 μm (see Fig. 4.7).

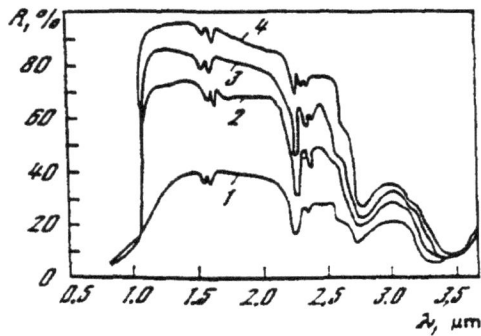

Figure 4.7:   Reflection coefficient of a transparent silicon solar cell manufactured in the USSR with a three-layer coating on the polished outer surface and different reflecting layers on the back surface without contact strips (depth of p–n junction is less than 0.5 μm): (1) nickel and titanium; (2) aluminum; (3) copper; (4) silver.

Comparably high reflection coefficients can be attained by another simple method, namely, by using the organosilicon compound (on the back surface of transparent cells) to attach glass coated with aluminum or silver. This procedure can be used to attach to the outer surface of the cell or group of cells a protective glass whose surface facing the element carries a grid of reflecting metal at points lying above the current contacts to the individual solar cells or above the electrical junctions between them. By varying the width of the reflecting grid lines, it is possible to adjust the temperature of such cells during an increase or a reduction in solar flux. Figure 4.8 illustrates the configuration of a module consisting of parallel-connected silicon cells, transparent in the infrared, with protective glass on both sides and a reflecting grid in the inner surface.

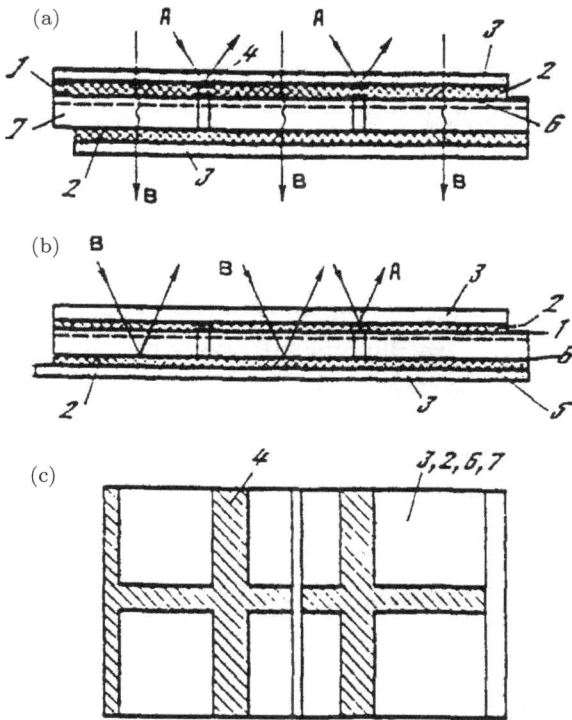

Figure 4.8: Modules of parallel-connected transparent solar cells: (a, b) with antireflective and reflective coating on the back surface, respectively; (c) top view. (1) Current contacts and intercell connections; (2) transparent organosilicon compound; (3) protective glass cover; (4) grid of aluminum or silver reflecting strips above the upper current contacts and connections; (5) reflective coating on the back glass; (6) antireflective coating; (7) solar cells; (A) solar radiation; (B) infrared solar radiation with $\lambda > 1.1\,\mu$m and $\lambda > 0.9\,\mu$m in the case of solar cells made from silicon and gallium arsenide, respectively.

Solar batteries consisting of such modules have a lower working temperature in space (by $25 - 35°$C) and greater thermal stability. This was confirmed experimentally over long periods of time in space when the batteries were carried on board the Venera-9 and Venera-10 automatic interplanetary stations [13].

It is important to note that the optical characteristics of transparent solar cells made from different semiconducting materials and carrying a reflecting back coating are very similar to the optical parameters of dichroic beam-dividing mirrors, which means that such cells may find important applications in high-efficiency photovoltaic systems using spectral subdivision of solar radiation and subsequent conversion into electric power by cells with different spectral sensitivities. Transparent solar cells then perform two functions simultaneously, namely, beam-splitting and active conversion.

4.2.5   *Solar cells with two-sided spectral sensitivity*

It was found that solar cells transparent in the infrared are sensitive to light when they are illuminated not only from the upper side but also from the back side. Solar cells that can generate electric current when illuminated on both sides are useful both in space and in terrestrial applications because the use of such cells improves the efficiency of solar batteries. Solar batteries consisting of cells with two-sided sensitivity can convert (in low-lying orbits) not only direct solar radiation, but also radiation reflected by the Earth, since the albedo of the Earth can reach 0.8–0.9 over portions of the orbit (under continuous cloud cover). This means that appreciably greater power levels can be generated. In terrestrial applications, these solar batteries can be provided with additional reflectors that illuminate the back surface of the two-sided cells, or batteries can be mounted on tall supports, so that radiation reflected by snow or sand reaches the usually unilluminated back surface.

Experiments have shown that when cells of the usual transparent design are illuminated from the back, the increase in the current and output power is no more than 10% or 20% of the initial values (this was confirmed, among other things, during the first few hours of the Venera-9 and Venera-10 missions), and the main advantage of transparent cells of conventional design is that the working temperature of the solar batteries is lower [13].

The use of two-sided cells with isotype junctions on the back surface in low-orbit satellites results in the availability of reserves of power. In this experiment, the solar irradiance intercepted by the back face of the two-sided cell was 0.3 of the irradiance on the upper, front surface because the average albedo of the Earth is close to this figure. Consequently, these measurements can be used to estimate the possible gain in power generated by solar batteries consisting of two-sided cells with isotype junctions on the back surface when they are mounted on low-orbit satellites (orbit altitudes of 200–400 km). These results were subsequently confirmed qualitatively by direct experiments in space [14]. The average albedo of the Earth during this flight was 0.25, and the current drawn from the two-sided solar batteries was, on average, 17–18% greater (15 ± 2% during the first ten orbits) than that delivered by one-sided solar batteries of conventional design.

4.2.6   *Space experiments*

Silicon solar cells, transparent in the infrared and provided with heat-reflecting mirror coatings, have a low equilibrium working temperature and can be widely used not only in batteries for automatic interplanetary stations, such as those in the Venera series [13], which operate under high levels of illumination, but also for studying direct solar radiation as well as the radiation reflected from the Earth and its cloud cover. This has been confirmed by space tests reported in [15].

Solar batteries mounted on low-orbit satellites (200–400 km) must be protected from overheating due to the Earth's albedo and its thermal emission. The Earth's albedo and, consequently, the thermal regime of solar batteries, are very much dependent on the cloud cover and the optical properties of the underlying surface in the particular locality. The amount of reflected solar radiation reaching the back surface of solar battery panels may, in turn, be determined by solar-cell probes facing the Earth (and shielded from direct solar light), provided their temperature in orbit can be determined by some independent method.

All these interrelated questions were investigated in experiments performed on the Cosmos satellites, in particular, Cosmos-1061, -1280, and -1301. These satellites were designed to investigate the natural resources of the Earth.

Two rectangular panels bearing small solar batteries and sensors were mounted inside the container carrying the scientific equipment of these satellites. The lower panel could be accurately oriented so that its normal pointed toward the center of the Earth, whereas the upper panel was at right angles to the direction of incidence of solar radiation at any time in orbit. Four small solar batteries were mounted on each panel to ensure the reliability and precision of the final results. Each battery consisted of 78 flat solar-cell modules of 54×40 mm and was similar to the solar batteries used on the automatic interplanetary stations Venera-9 and Venera-10 [13]. The solar cells and the supporting panel were transparent to the solar infrared in the wavelength range between 1.1 and 2.5 μm, and were shielded from cosmic rays and the solar ultraviolet by thin radiation–shielding glass filters attached to the antireflective surface of the silicon solar cell by transparent silicon rubber (the optical properties of the glass filters and the rubber were found to be stable in the course of prolonged tests in space). Heat-reflecting aluminum coatings were deposited on the inner face of the shielding glass (above contact junctions between the cells) (see Fig. 4.8(c)). Glass filters attached to the rear of the modules were also coated with a heat-shielding aluminum layer (see Fig. 4.8(b)) which reflected not only rays with wavelengths between 1.1 and 2.5 μm, transmitted by the transparent solar cells, but also 83–85% of the solar spectrum (0.3–2.5 μm) incident on the rear surface after reflection from the Earth and its cloud cover. Two batteries on each panel were operated in the short-circuit current mode, and the current was used to determine the amount of solar radiation incident on the Earth (panel 2) and reflected from the Earth (panel 1). The open-circuit voltage of the other batteries was used to determine the panel temperature (the temperature dependence of this voltage was determined in laboratory experiments).

All the small solar batteries used in these experiments had very similar short-circuit currents, carefully measured under AM0 sun simulators in the laboratory. An initial reference point was thus obtained (the short-circuit current of batteries under 1360 W/m^2), and was used to determine the flux density of radiation reflected from the Earth at any time in orbit. Before the launching, all cells and batteries were carefully tested (and this procedure is also used at present) by special laboratory methods without sun simulators [16, 17].

Figures 4.9(a) and 4.9(b) show the time distribution of solar radiation reflected from the Earth, which is typical for low-lying orbits.

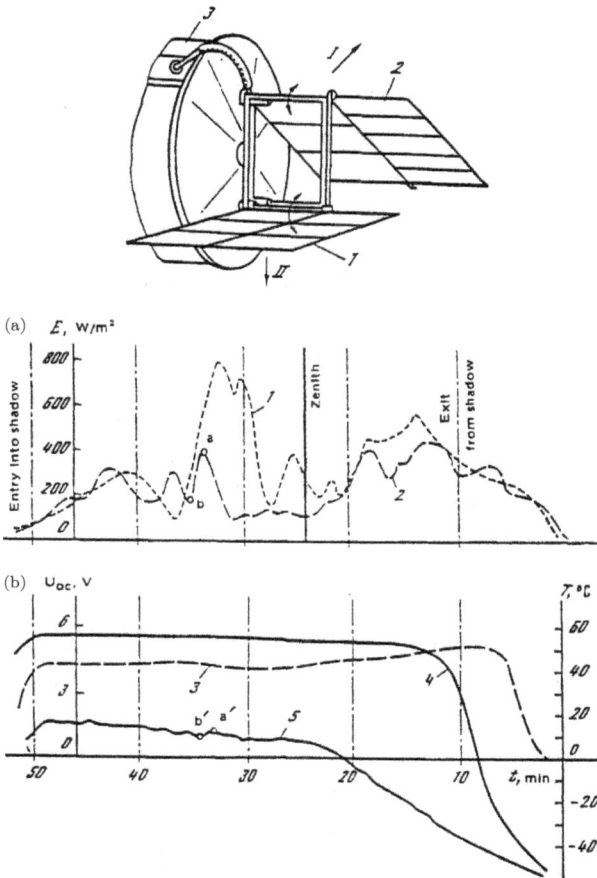

Figure 4.9: Solar radiation flux density reflected from the Earth and from its cloud cover, measured by one of the Cosmos satellites in orbit (a), and variation in the open-circuit voltage and temperature of solar batteries (b): (1, 2) the first and second orbits, respectively; (3) open-circuit voltage generated by batteries facing the Sun; (4, 5) temperature of batteries facing the Sun and the Earth, respectively. Also sketched are dispositions of flat panels carrying small solar batteries on some of the Cosmos satellites: (1, 2) panels facing the Earth and Sun, respectively; (3) container housing scientific equipment; (I) direction of the Sun; (II) direction of the Earth

These data were obtained by one of the Cosmos satellites in an orbit with minimum duration of illuminated segment (about 52 min). This type of orbit requires that a higher output power be generated by the solar batteries, and the radiation reflected by the Earth is the "reserve" necessary for generating additional power. The orbital planes characterized by minimum illuminated segments lie in the plane of incidence of the light flux. The reflected flux density usually rises on either side of the zenith line because there is then a sharp improvement in reflection by the cloud cover and in the incidence of reflected flux on the back of the panels.

The maximum temperature of the panels is low (it does not exceed 57–58°C, in contrast to the 70–75°C typical for solar cells of conventional design [11]) because the cells are transparent to the solar infrared and heat-reflecting coatings are employed. It is clear from Fig. 4.9(b) that the temperature of panel 1 is a function of the amount of radiation incident upon it after reflection from the Earth (points a′ and b′ on curve 4 correspond to points a and b on curve 2).

The above studies have demonstrated that the working temperature of solar cells and batteries can be substantially reduced by using heat-reflecting mirror coatings and solar cells which are transparent to the solar infrared.

Semiconductor solar cells that convert solar radiation directly into electric power can be used not only as the sources of power for spacecraft, but also as sensors of direct and reflected light when the Earth is investigated from space, and in studying the transparency of the atmosphere of other planets, for example, Venus. Most of the spacecraft entering the Venusian atmosphere were therefore equipped with small solar batteries calibrated on the Earth under laboratory conditions. The information provided by these batteries not only complemented the spectral measurements of the transparency of the Venusian atmosphere, but was also used to estimate the diffuse fraction of the light flux in that atmosphere. During the descent of Venera-5 on 16 May 1969, bursts of illumination were recorded for the first time on the night side of this planet, which together with the light measurements performed by Venera-9 and Venera-10, and measurements of the radio noise due to electrical discharges on Venera-11 and Venera-12, led to the conclusion that there were lightning discharges in the atmosphere of the planet.

Small batteries were developed to withstand a load of $300g$ (g is the gravitational acceleration) as they enter the atmosphere of the planet and also high temperatures and pressures (up to $300°C$ and 10^7 Pa). As a rule, each battery consists of 28–30 individual calibrated silicon solar cells connected in series with one another and with a load resistor producing an output voltage of 0–6 V when the solar flux density varies from 50 to 1500 W/m^2.

The small solar battery and other equipment on the Venera station were cooled down to about $-15°C$ prior to descent, so that an acceptable temperature could be established for the solar cells on most of the flight paths. A temperature correction was, therefore, introduced in the analysis of the results.

On 25 December 1978, the descending Venera-11 carried three small solar batteries mounted on its exterior. They were used to measure the solar flux density in the atmosphere of Venus, and showed that the flux density in the Venusian atmosphere was quite high (Fig. 4.10). This agrees with the results of calculations and other tests.

The uncertainties indicated by the vertical bars in Fig. 4.10 are very small, so that the flux is almost completely diffuse in character, and there is no direct component of solar radiation at these altitudes in the Venusian atmosphere. Similar experiments were performed, and results analogous to those shown in Fig. 4.10 were recorded during the descent of Venera-13 and Venera-14 in March 1982.

It has been shown above that GaAs cells, silicon cells transparent in the infrared and heat-reflecting coatings are important components of Soviet and Russian space programs.

These parts of space solar batteries have been successfully tested during the flight of the Soviet automatic interplanetary stations in the Venera series and during the operation of Lunokhod-1 and Lunokhod-2 on the Moon.

The fundamental point underlying these tests was that neither program could have been carried out with silicon solar cell batteries of conventional design and that they required the development of

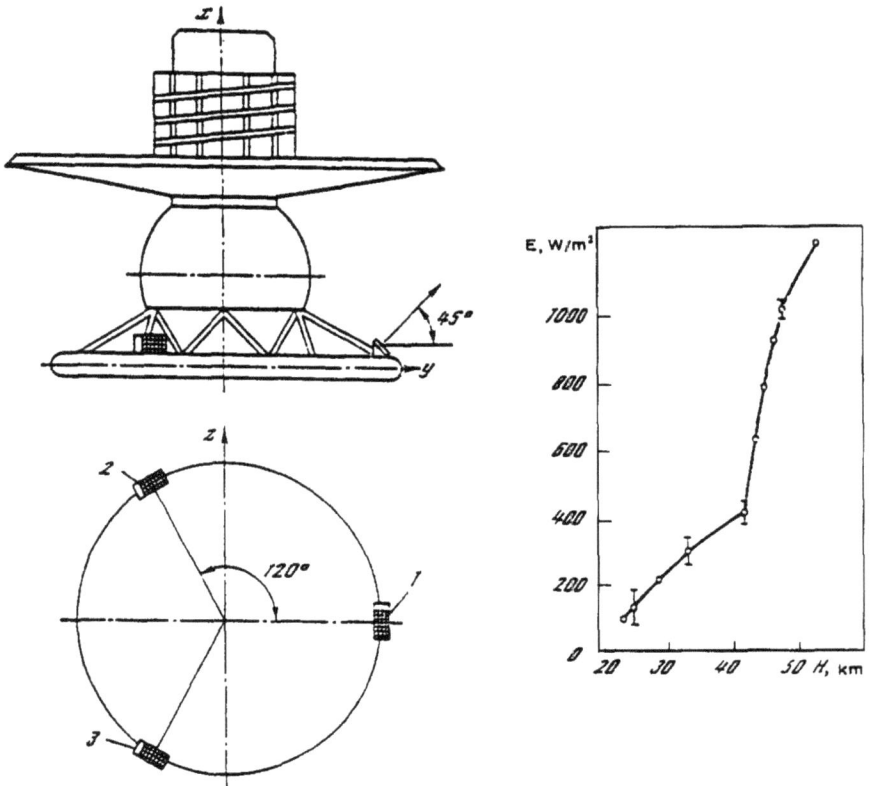

Figure 4.10: Solar flux density as a function of the height of the descending interplanetary station above the surface of Venus, obtained with calibrated silicon solar batteries. Also sketched is the disposition of small solar batteries (1–3) on the exterior of the Venera-11 interplanetary station.

new types and designs of silicon cells and solar batteries made of other semiconductor materials.

The solar-cell batteries used in these missions experienced increasing amounts of solar radiation and a simultaneous rise in the working temperature (in the case of missions to Venus) and a substantial increase in the equilibrium working temperature under constant solar illumination (in the case of the Lunokhod vehicles designed for prolonged operation on the lunar surface) as compared with equipment working near the Earth. Calculations have shown that the equilibrium working temperature of the Lunokhod solar batteries illuminated by the Sun and heated by the substantial thermal radiation of the Moon was $125 - 145°C$. The temperature of solar batteries consisting of conventional silicon solar cells over the flight path from the Earth to Venus increased gradually from $65°C$ to $150°C$.

Passive temperature control of the kind for inclined panels that are not accurately oriented to face the Sun, or a substantial reduction in the solar panel occupation coefficient, could not be used to reduce the working temperature because the size and design of the probes were subject to stringent restrictions. It was, therefore, decided to use silicon cells transparent to the infrared as a basis for the solar batteries on Venera-9 and Venera-10.

The lower equilibrium temperature and improved operation of solar batteries under the increased concentration of solar radiation (by factors between 2 and 8) were achieved by reducing the α_s/ε of the area covered by current contacts and of the area of solar panels that was free from photovoltaic cells. Stable low α_s/ε (less than 0.2) selective coatings were developed for the free surfaces of the panels (instead of the white enamels, which are known to darken after exposure to the solar ultraviolet). These coatings were based on films of radiation-resisting glass, whose rear surface was coated with aluminum or silver. The coatings were also successfully used to protect the cooling fins of Lunokhod-1 and Lunokhod-2 from overheating.

A grid of a highly reflecting metal (aluminum or silver) was deposited in a high vacuum (before being attached to the modules) on the rear surface of the protective cover glass on several large photovoltaic cells in order to achieve an equally low value of α_s/ε on cell areas occupied by contacts. This reflecting mosaic was produced by evaporating aluminum or silver through masks so that the configuration of the grid was the same as that of the contacts and the gaps between them. Because of the presence of the reflecting grid on the back of the cover glass, the area occupied by the contacts reflected 84% (in the case of aluminum) or 92–94% (in the case of silver) of the solar radiation. The solar absorption coefficient of areas occupied by the contacts was, at the same time, reduced from 0.75 to 0.16 or even 0.06–0.08, whilst the hemispherical emission coefficient remained high at 0.86 (because of the outer layer of glass), both for the surface of the silicon cells and for the surface of the contacts. The equilibrium temperature of the contact was thus held low, and the transfer of heat from the heated semiconductor surface to the cooled contacts resulted in a reduction in the mean equilibrium temperature of the solar batteries. The area occupied by the contacts and the gaps between them in the parallel module could be varied within broad limits. If the solar battery operates with enhanced concentration (when it is mounted on a glass grid or a polymer film in order to improve its limiting characteristics and thermal stability), a change in the area of the heat-reflecting contacts is the only way of achieving the design temperature since the equilibrium temperature of solar batteries can readily be evaluated on the basis of thermophysical calculations.

Parallel modules of silicon cells which are transparent to the solar infrared, and heat-reflecting coatings of aluminum deposited in vacuum on 10–12% of the inner surface of the cover glasses (radiation-stable glass, 170 or 300 μm thick), were used for the solar batteries carried by Venera-9 and Venera-10. The advantage of using a single cover glass for several cells at once (a module consists of between four and six cells of 5 cm^2 each) is also that the reliability of the radiation shield is improved (see Fig. 4.8).

Calculations have shown that the equilibrium temperature of such solar batteries should be lower by $30 - 35°$ C over the entire route between the Earth and Venus as compared with solar batteries of conventional design.

Thus, the use of cells that are transparent to the solar infrared, and of heat-reflecting coatings deposited on the inner side of cover glasses on areas above the contacts, has ensured a favorable temperature distribution and improved reliability of solar batteries on Venera-9 and Venera-10.

Several months of operation in space have shown that the active elements and coatings of the new space solar batteries operated satisfactorily throughout that time.

It is clear from published information [2, 3] that for equal initial efficiencies (at room temperature), the gallium arsenide cells generate substantially more electric power than silicon cells at $130 - 140°$ C, which is the temperature expected on the basis of thermophysical calculations at which the Lunokhod solar batteries were to operate. The development of high-efficiency gallium arsenide cells has resulted in Lunokhod-1 and Lunokhod-2 solar batteries of a relatively large area (about $3\,m^2$). These are made entirely from this semiconducting material and were used in the first tests of gallium arsenide solar cells not only in space, but also on the lunar surface in the presence of dust, micrometeorites, and considerable temperature gradients [13].

The Lunokhod solar battery provided all the electric power not only for the moon buggy but also for all the equipment carried between the Earth and the Moon. It also provided the power supplies in lunar orbit prior to descent onto the lunar surface.

It is important to note that the battery on Lunokhod-1 (delivered to the lunar surface on 17 November 1970) operated successfully for 10 lunar days (up to 16 October 1971). The working current delivered by the battery was found to fall by about 6% over that period. The gallium arsenide solar batteries in Lunokhod-2 (delivered to the lunar surface on 16 November 1973) operated successfully practically without deterioration in their working characteristics for the entire design program (5 lunar days). Now, all these components — new solar cells and original selective coatings — are widely used in the modern Russian space programs.

4.2.7 *Terrestrial tests*

At first, it was the opinion that establishing terrestrial photo-generators from various semiconducting solar cells would meet no serious scientific, technological, or engineering difficulties because the operating conditions for this type of equipment are much less severe, and repair opportunities are much higher than those for space solar batteries. However, it was discovered rather soon that the distorting effect of, for example, humidity on solar cell parameters is often stronger than that of temperature cycling in orbit. This makes the problem of terrestrial solar cell hermetizing one of the most important problems. It was realized that it is necessary to protect not only the solar cells themselves, but also other units such as additional reflectors or solar concentrators.

Terrestrial solar cell coatings are designed to protect not the separate solar cells, but the whole module. The solar cell coating is hermetically attached to the cell surface so as to make the cells water-resistant.

In the simplest case, shielding is provided by using an optically transparent hermetizing compound or by putting the cells under a cover which also protects the contacts.

The initial characteristics of the solar cells can, unfortunately, deteriorate during their use even if the cells are protected with hermetizing coating.

A rise in temperature usually leads to an increase in photoelectric current and a decrease in the electromotive force, output power, and efficiency of the solar cells. The gradient of the decline in output power depends on the nature of the semiconducting material; it is low for wide-gap materials and high for narrow-gap ones. For silicon solar cells, a rise of temperature by $100°$ C leads to a decrease in their output power by 45%, and for gallium arsenide cells, by 25%. (The width of the band gap for Si is $1.02\,\text{eV}$ and for GaAs, $1.43\,\text{eV}$.)

A rise in density of the incident radiation flux by several times can lead to a drastic fall in output power if the series resistance of the solar cells is high; about 1 Ω/cm^2 series resistance of conventional solar cells amounts to about 0.5–0.6 Ω/cm^2, and they can operate (without deterioration) under the conditions of a 5–7-fold increase in solar radiation density, which is common for the conditions of the middle part of Russia (usually about 400–800 W/m^2).

There are several ways of lowering the series resistance: for example, by attaching a dense contact grid to the upper side of the cells (with optimized strip size and distances between them), which not only reduces the series resistance but also allows one to use more effectively an increased density of radiation flux, produced usually by solar concentrators of various types. It has been shown in several papers that sufficient increase in the density of the incident solar flux leads to a growth of solar cell efficiency, owing to useful electric drift fields in the semiconductor's bulk (provided there is no fall in output power due to current dissipation). Experiments verified this conclusion [5]. With a decrease in the series resistance of solar cells to 0.1 Ω/cm^2, the maximum efficiency of energy conversion was observed under a 40–50-fold intensity of solar radiation. With a decrease in series resistance to 0.01 Ω/cm^2 or less, it is possible to convert effectively into electricity solar radiation exceeding one sun solar flux by 500–700 times.

It is necessary to emphasize that in this manner the previously mentioned adverse effects of increased solar flux on cell efficiency can be overcome (provided there is no overheating of the cells during use: the limiting temperature being the operation limit of the contact layers and coatings, usually about 150–200°C).

Both in space and on the Earth, solar cells and semiconducting materials run the danger of irreversible changes. A particularly large fall in the output power of solar cells and batteries can be observed due to corpuscular radiation — protons and electrons — in the Earth's radiation belts, and due to repeated temperature cycling while entering the Earth's shadow and leaving it. High-temperature-induced stresses arise inside the solar cells under such cycling because of the difference in thermal expansion coefficient among the various layers of the semiconductor (in the case of hetero- and homojunctions), contacts, and antireflective and protective coatings. Such stresses lead to the mechanical destruction of a solar cell if they exceed the strength of the layers or the forces that keep them together.

In many countries, sufficient experience has accumulated in developing solar cells and batteries to render them capable of resisting the negative effects of radiation and temperature cycling. Nowadays, solar batteries can operate in space and on the Earth for dozens of years without any appreciable decrease in output power.

This difficult problem has been solved thanks to a better understanding of the complex physical processes occurring in semiconducting materials as well as at the interfaces with the other materials forming the solar cell, and which cause the degradation of parameters under exposure to various external factors. These processes are investigated in the laboratory with the help of various up-to-date techniques for analyzing the composition, impurities and defects of the materials. These include electron and optical focused-beam microscopy, ion-induced spectroscopy, Auger spectroscopy, mass spectrometry, X-ray microanalysis, photo-, cathode-, and electroluminescence spectroscopy, etc. [16, 17].

Two more types of effect attracted the attention of researchers. The first one causes a degradation which may be termed "chemico-thermal". The second one is "photon" degradation.

Chemico-thermal degradation appears, for example, because of the influence of a spacecraft's residual atmosphere and engine exhaust gases on the solar cell parameters. The air in large cities, polluted with gaseous alkaline and acidic exhausts, is also very harmful to solar cells in terrestrial applications and to their optical coatings. Unusual chemical reactions of free radicals occurring on the unprotected surfaces of solar cells, under increased temperature, cause short circuiting in p–n junctions, contact corrosion, and coating darkening.

The effect of photon degradation was not discovered at once for it is rather difficult to distinguish it from the effect of corpuscular radiation and chemico-thermal degradation. For a long time, it was the opinion that the destructive effect of solar radiation on solar cells could manifest itself only in darkening of the optical coatings. The development of light-resistant multiple layer coatings, where the upper layer is a thin glass plate with CeO_2 additive (which absorbs ultraviolet radiation of wavelengths less than 0.36 μm), had the effect of reducing the degradation in the optical properties of the coatings to rather small values (0.5–2.5%) — even under continuous use aboard space vehicles for several years.

Having effectively solved the coating degradation problem, it was rather unexpected for many researchers to discover that the visible part of the solar spectrum also causes deterioration in the optical properties of solar cells. In the course of early experiments, when the combined effects of solar light, corpuscular radiation, and temperature were investigated, some important peculiarities of the simultaneous effect of several damaging factors were detected. Such experiments simulate rather well the real conditions of solar cells both in space and on the Earth.

There are several ways to decrease photon degradation in solar cells, such as by avoiding Ag atoms in the silicon base layer, eliminating the mechanically damaged surface layer before diffusion, and carrying out the diffusion of dopant atoms at temperatures not exceeding 875° C. For example, solar cells for which, in the process of their fabrication, the diffusion process was carried out at a temperature of 950° C were observed to undergo photon degradation of 3–6% (while illuminated by a sun simulator with a radiation flux density equal to 1000 W/m²). At a diffusion temperature of 900° C, the observed degradation was 1–3% and at 875° C, only 0.5%.

It is especially important to take into account the effect of photon degradation when fabricating standard solar cells for the adjustment of sun simulators. Such cells must obviously have stable characteristics.

There is no doubt that the discovered new types of solar cell degradation will be studied thoroughly, and ways will be found for their prevention. At such time, solar cells will be highly effective, stable and reliable sources of power, usefully converting solar energy into convenient electricity.

4.2.8 *Final remarks*

(1) The need for reliable and effective power sources for spacecraft was a strong impulse for developing, in the course of the Soviet space program, photovoltaic solar radiation converters of various semiconducting materials, primarily monocrystalline silicon and gallium arsenide.

(2) The necessity of protecting solar cells and batteries from the effects of radiation, overheating and temperature cycling led to new designs for solar batteries and the development of various protective optical coatings.

(3) Tests carried out in the laboratory and in space for 20–30 years helped to produce a better understanding of the nature of the main types of degradation of these power sources and to develop ways of overcoming them.

(4) Simultaneously with laboratory measurements and space tests, experiments on the terrestrial applications of solar cells were carried out in several climatic regions of the former USSR. More than 200 batteries of low output power (0.2–0.3 kW) are operating successfully in Russia and in the various republics of the former USSR, from the south to the Arctic Ocean and Lake Baikal.

(5) Political and economic changes taking place in Russia speak in favor of more intensive international cooperation both in the sphere of fundamental research in the field of semiconductor physics, and in the development of international projects on solar photovoltaic power plants and new technologies for solar cell and battery fabrication.

4.2.9 *References*

[1] A. F. Ioffe, *The Physics of Semiconductors* (USSR Academy of Sciences, Moscow–Leningrad, 1957, English translation by Academic Press, New York, 1961).

[2] H. S. Rauschenbach, *Solar Array Design Handbook* (The Principles and Technology of Photovoltaic Energy Conversion) (Van Nostrand Reinhold, New York, 1980).

[3] A. L. Fahrenbruch and R.H. Bube, *Fundamentals of Solar Energy Conversion* (Academic Press, New York, 1983).

[4] M. M. Koltun, *Selective Optical Surfaces for Solar Energy Converters* (Nauka,Moscow, 1979, English translation by Allerton Press, New York, 1981).

[5] M. M. Koltun, *Solar Cells, their Optics and Metrology* (Nauka, Moscow, 1985, English translation by Allerton Press, New York, 1988).

[6] A. M. Vasil'ev and A. P. Landsman, *Semiconductor Photoconverters* (Soviet Radio, Moscow, 1971, in Russian).

[7] M. Evdokimov, M. B. Kagan, M. M. Koltun and A. Cherkasskii, Solar batteries. *Direct Conversion of Thermal and Chemical Energy into Electric Power* (VINITI, Moscow, 1977, in Russian).

[8] J. Mandelkorn and J. H. Lamneck, "New electric field effect in silicon solar cells", *J. Appl. Phys.* 44 1973, 4785–4791.

[9] L. B. Kreinin and G. M. Grigor'eva, "Solar batteries under bombardment by cosmic radiation", *Studies of Cosmic Space*, Vol. 13 (VINITI, Moscow, 1979, in Russian).

[10] M. B. Kagan, M. M. Koltun, A. P. Landsman, and T. L. "Lubachevskaya, Possible designs for cascade solar cells", *Geliotechnika (Applied Solar Energy*—in English translation) 1 1968, 7–20.

[11] M. M. Koltun and A. P. Landsman, "Semiconductor photoconverters transparent in the solar infrared", *Zh. Prikl. Spektrosk.* 15(4) 1971, 753–755.

[12] W. Palz, J. Besson, T. Hguyen Duy, and J. Vedel, "Review of CdS solar cell activities", I *Proceedings of the 10th IEEE Photovoltaic Specialists Conference*, Palo Alto, CA, 1973, pp. 69–76.

[13] G. S. Daletskii, M. B. Kagan, M. M. Koltun, and V. M. Kuznetsov, "Development of solar batteries for the Venera-9 and Venera-10 automatic interplanetary stations and the Lunokhod program", *Geliotechnika* 4 1979, 3–9.

[14] G. A. Boltyanskii, N. M. Bordina *et al.*, "The experimental two-sided solar battery for the Salyut-5 orbital station:, *Kosmich. Issled.* 18(5) 1980, 812–814.

[15] G. S. Daletskii, V. A. Karpukhin, M. M. Koltun, and V. M. Kuznetsov, "Experimental study of the Earth and its cloud cover on the heat balance of solar cells on one of the Cosmos satellites", *Geliotechnika* 5 1982, 3–6.

[16] M. M. Koltun, Yu. M. Kuznetsov, E. l. Ran, G. V. Sasov, G. V. Spivak, and N. M. Khvastunova, "Scanning electron microscopy of silicon photoconverters", *Poverkhnost (Fiz., Khim., Mekh.) (Surface Physics, Chemistry, Mechanics)* 1(10) 1982, 70–79 (in Russian).

[17] M. I. Gercenstein and M. M. Koltun, "Electroluminescent introscopy of solar cells",. *Sol. Energy Mater. Sol. Cells* 28(1) 1992, 1–7.

4.3   Discussion following Mark Koltun's presentation

Appelbaum: You did not mention the use of indium phosphide or light-weight amorphous silicon cells for space.

Koltun: In recent years, most of the effort was put into crystalline silicon and gallium arsenide solar cells for use in space, but indium phosphide is a potentially useful material because its resistance to ionizing radiation is very high. We made some indium phosphide cells with efficiencies of about 16% and tested a few samples on space flights but they have not been used in large quantities. As for amorphous silicon, it was not a part of the Soviet space program, so research on this kind of solar cell never achieved the attention it did in countries such as Japan, for example. Unfortunately, with the break-up of the Soviet Union, much of the coordinated effort has now stopped. Indium phosphide cells were made in Kishinev, which is now part of the independent state of Moldavia. Silicon ingots, on the other hand, were manufactured in the Ukraine and not in Russia. So it is difficult to see how progress will be made in the current political situation.

Roy: What about polycrystalline solar cells?

Koltun: This material is not stable in space and so, like amorphous silicon, it did not receive much attention in the Soviet Union.

Carmel: Reliability is a key feature for grid-connected photovoltaic systems. According to your experience, what would be the degradation in module efficiency after 10 years of use?

Koltun: If the modules are well made, the degradation should not be more than 5% and, as you know, we have much experience in this area. But I must emphasize that technology is the crucial issue. Many European and American firms use silver contact strips and polymer glues and are only prepared to guarantee their modules for 10 years. I believe, however, that one could use space technology in a cost-effective manner for the electrical contacts and, in this way, produce modules that would last 25–30 years with negligible degradation.

Elazari: What is the present cost of photovoltaic modules in Russia?

Koltun: Two years ago, the prices — if expressed in hard currency — were unrealistically low, so I shall not quote what they were. At present, there are a number of small companies producing cells at somewhat below American or European prices: about $4/W for the cells and $7–$8/W for installations. Western prices are currently about $10–$12/W for installations.

Roy: As a historical curiosity, did the first Sputnik satellite use solar cells?

Koltun: No. Only on the 3rd Soviet Earth satellite was this technology first employed — the same year as Vanguard, the first American space vehicle to use solar cells. Our solar cell research program was started by Arkady Lantzman (whose graduate student I was), based on the semiconductor research of academician Yoffe, but it was forbidden to publish any of that work in the 1960s.

Hallak: How did you make the physical connection between the two parts of your first cascade cells?

Koltun: We employed simple mechanical bonding.

Roy: Are you optimistic about the future of cascade cells for terrestrial uses?

Koltun: Very much so. Boeing Corp. recently succeeded in producing a 38% efficient cascade cell. This was, of course, a laboratory sample but in the future such efficiencies may be mass-producible.

Chapter 5

1994

5.1 Editor's Foreword

The two keynote speakers at the *6th SedeBoqer Symposium*, in which photovoltaic presentations dominated, were the late Stephen Kaneff of the Australian National University and Michael Grätzel of the École Polytechnique Fédéral de Lausanne. Prof. Kaneff, the designer of the then world's largest paraboloidal solar concentrator ($400\,m^2$), similar to the one that was later erected at SedeBoqer, spoke of the potential of such dishes for photovoltaic and other purposes. The great attraction of a large dish concentrator for solar photovoltaic purposes is that a 100 kW (for example) photovoltaic module placed at its focus is simpler and of potentially lower cost than a solar-thermal receiver of similar power.

Prof. Grätzel presented an overview of his exciting recent invention, the "nanocell" — better known to the world as the "Grätzel cell". For an audience which at that time was familiar mainly with the physics of solid-state junction cells, the biggest surprise was that these novel devices contain no built-in electric field. Both presentations together with the vigorous discussions that followed were included in the symposium proceedings and are reproduced here.

5.2   Low-cost and efficient photovoltaic conversion by Nanocrystalline solar cells (Prof. Michael Grätzel)

Keynote presentation by Michael Grätzel, Institut de chimie physique, École Polytechnique Fédéral de Lausanne, Ch-1015 Lausanne, Suisse

5.2.1   *Abstract*

The quality of human life depends to a large degree on the availability of energy sources. The present worldwide energy consumption exceeds already the level of 6000 GW. It is expected to further increase sharply from the rising demand of energy in the developing countries. This implies enhanced depletion of fossil fuel reserves, leading to further aggravation of the environmental pollution exerting adverse effects on the wellbeing of mankind. Adding the dangers arising from the accumulation of plutonium fission products from nuclear reactors, the quality of life on Earth is threatened unless renewable energy resources can be developed in the near future. Photovoltaic solar energy converters are expected to make important contributions to identify environmentally friendly solutions of the energy problem. One attractive strategy discussed in this lecture is the development of systems that mimic natural photosynthesis in the conversion of solar energy for the fixation of carbon dioxide. The task to be accomplished by these systems is to harvest sunlight to produce electricity. Learning from the concepts used by green plants, we have developed a molecular photovoltaic device whose overall efficiency for solar energy conversion to electricity has already attained 10%. The system is based on the sensitization of nanocrystalline films by transition metal charge-transfer sensitizers. In analogy to photosynthesis, the new chemical solar cell achieves the separation of the light absorption and charge carrier transport processes. Extraordinary yields for the conversion of incident photons into electric current are obtained, exceeding 90% for transition metal complexes within the wavelength range of their absorption band. Conventional photovoltaic cells for solar energy conversion into electricity are solid-state devices and do not economically compete for base load utility electricity production. The low cost and ease of production of the new nanocrystalline cell should benefit large-scale applications, in particular, in underdeveloped or developing countries. These regions of the Earth benefit from generous sunshine, rendering the availability of a cheap solar cell technology important in view of improving the quality of life and preserving natural resources. Quite aside from its intrinsic merits as a photovoltaic device, the nanocrystalline films development opens up a whole number of additional avenues for energy storage, ranging from intercalation batteries to the formation of chemical fuels. These nanocrystalline systems will undoubtedly promote the acceptance of renewable energy technologies, not least by setting new standards of convenience and economy.

5.2.2 *Introduction*

There can be no question that the quality of human life is intimately associated with the ready availability of energy resources. At present, the world's energy consumption rate exceeds already the stunning figure of 6000 GW. This is expected to grow rapidly in the next decades due to the increase in demand from the developing countries. As these countries strive to achieve a higher standard of living, their energy consumption will augment dramatically. The overwhelming part of our supply arises from the chemical energy stored in fossil fuels. These reserves are therefore being rapidly depleted. Furthermore, their combustion has led to unacceptable levels of pollution of our environment. Further acceleration of this process would lead to disastrous climatic consequences and severe deterioration of the quality of life on the Earth. Nuclear energy does not provide an acceptable alternative. Apart from the issues of safety and radioactive pollution, these reactors have generated already a large amount of plutonium products, exceeding 700 tons whose storage generates a severe security problem. The mere facts that about 10 kg of plutonium are sufficient to produce a nuclear bomb and the lethal dose for humans is in the microgram range render perfectly clear the terrifying consequences the irresponsible use of this waste could impose on us and many future generations to come.

It is evident from this analysis that the quality of life on the Earth is threatened unless renewable energy resources can be developed in the near future. In this context, I would like to quote the eminent French scientist Frederic Joliot Curie (1900–1958), who in 1956 made the following statement in front of the Economic Council of the French Republic: "A mon avis, il faut s'occuper très sérieusement, et dès maintenant de l'énergie solaire. C'est sans doute par des recherches techniques que l'onarrivera à améliorer considérablement les procédés d'utilisation de l'énergie de la radiation solaire. C'est je lerépète, un problème d'unetrès grande importance, qui devrait intéresser l'industrie française et les établissements de recherches appliquées de l'Etat...".

Photovoltaics is expected to make decisive contributions to identify environmentally friendly solutions to the energy problem. In a conventional p–n-junction photovoltaic cell made, for example, from silicon, the semiconductor assumes two roles at the same time: it harvests the incident sunlight and conducts the charge carriers produced under light excitation. In order to function with a good efficiency, the photons have to be absorbed close to the p–n interface. Electron–hole pairs produced away from the junction must diffuse to the p–n contact where the local electrostatic field separates the charges. In order to avoid charge carrier recombination during the diffusion, the concentration of defects in the solid must be small. This imposes severe requirements on the purity of the semiconductor material, rendering solid-state devices of the conventional type very expensive. Molecular photovoltaic systems separate the functions of light absorption and carrier transport. The light harvesting is carried out by a sensitizer which initiates electron transfer events leading to charge separation. This renders unnecessary the use of expensive solid-state components in the system. While simple from the conceptual point of view, the practical implementation of such devices must overcome formidable obstacles if the goal is to develop molecular systems that convert sunlight to electricity with an efficiency comparable to silicon cells and to meet the stability criteria for practical applications. Our success in bringing this project close to commercial maturity owes much to the recent progress in the molecular engineering of stable and efficient charge-transfer sensitizers based on platinum metals, that is, ruthenium and osmium. The approach taken by us will now be outlined in more detail.

5.2.3 *Light harvesting by transition metal complexes anchored to nanocrystalline oxide films*

The fundamental processes involved in any photovoltaic conversion process are as follows:

(i) the absorption of sunlight,
(ii) the generation of electric charges by light,
(iii) the collection of charge carriers to produce electricity.

In any case, the components of the system must be selected to satisfy the high stability requirements encountered in practical applications. A photovoltaic device must remain serviceable for 20 years without significant loss of performance. One of the most remarkable achievements of research in inorganic chemistry during the last two decades has been the development of a great variety of platinum metal complexes [1], mainly of the elements rhenium, rhodium, osmium and ruthenium, many of which are very stable and display good absorption in the visible. Our work has focused on the molecular engineering of suitable ruthenium and osmium compounds. Bis(bipyridyl)Ru(II) complexes having the general formula *cis*-X_2bis(2,2'-bipyridyl)-4,4'-dicarboxylate)-ruthenium(II), where X = CI^-, Br^-, I^-, CN^-, and SCN^-, exhibit the most promising properties so far. We performed a systematic study of their luminescence, visible light absorption, electrochemical and photoredox behavior [2]. Among these compounds, *cis*-di(thiocyanato)bis(2,2'-bipyridyl)-4,4'-dicarboxylate)-ruthenium(II), (I), performs as an outstanding solar light absorber and charge-transfer sensitizer, unmatched by any other dyestuff known so far. Its broad range of visible light absorption and relatively long-lived excited state render it an attractive sensitizer for homogeneous and heterogeneous redox reactions. The structural features of this complex are shown in Fig. 5.1.

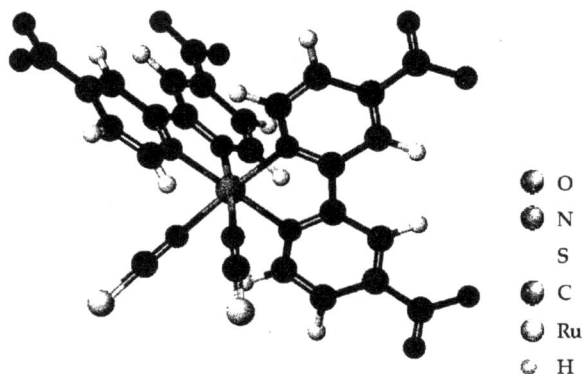

Figure 5.1: Structure of *cis*-di(thiocyanato)bis(2,2'-bipyridyl)-4,4'-dicarboxy-late)-ruthenium(II), (I), which shows outstanding properties as a charge-transfer sensitizer for nanocrystalline TiO_2 films.

In our photovoltaic system, the ruthenium complex is anchored to the surface of a wide-band oxide semiconductor, such as TiO_2. Due to its large band gap (3.2 eV), the semiconductor does not absorb visible radiation. Light harvesting is carried out by the ruthenium complex acting as a sensitizer. For a monolayer of dye adsorbed on a flat surface, there is the notorious problem of insufficient light absorption. The situation is depicted schematically in Fig. 5.2. On a smooth surface *cis*-di(thiocyanato)bis(2,2'-bipyridyl)-4,4'-dicarboxylate)-ruthenium(II) absorbs less than 1% of the incident light even in the wavelength range of maximum absorption, that is, 530 nm. One could naturally think, then, of depositing several molecular layers of sensitizer on the semiconductor membrane in order to increase the light absorption. This would however be a mistaken tactic since the outer dye layers would act only as a light filter, with no contribution to electrical current generation. The application of a monomolecular layer of sensitizer is therefore unavoidable.

A solution to the problem of light harvesting through such extremely thin molecular layers was found by us in the application of nanocrystalline films [3]. It is possible using the sol–gel method to produce transparent films consisting of colloidal titanium dioxide

Problem of the absorption of light by a monomolecular sensitizer layer

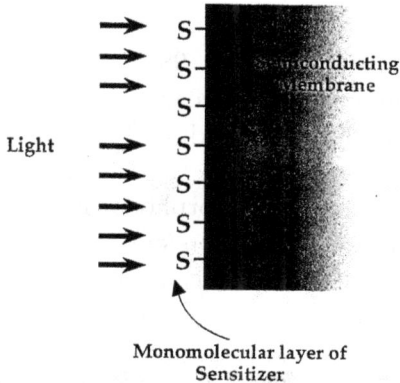

Monomolecular layer of Sensitizer

The absorption is insufficient if the membrane is smooth!

$$\eta_{abs} = 1 - 10^{-\Gamma_s \, \sigma_s} \approx 0,0005 \to 0,05$$

Γ_s number of sensitizer molecules per cm^2 of area ($\approx 10^{14}$ cm^{-2})

σ_s absorption cross section for light

Solution to this problem : Use surfaces of **high roughness**

Figure 5.2: Light harvesting by monomolecular films of sensitizers.

particles with diameters of 10–30 nm. The electronic contact between the particles is produced by a brief sintering at about 500° C. A nanoporous structure with a very high internal surface area is thereby formed. For example, the effective surface of an 8 micron thick layer of such a colloidal structure is about 720 times greater than that of a smooth film. On the geometric projection of such a surface, a sensitizer concentration of $\Gamma = 1.2 \times 10^{-7}$ moles/cm^2 is reached. The optical density

$$OD(\lambda) = \Gamma \times \varepsilon(\lambda) \tag{5.1}$$

calculated for this coating level is 1.68 using $\varepsilon = 1.4 \times 10^7 \, cm^2/mole$ for the extinction coefficient, which is the value obtained for complex I at 536 nm. The light harvesting efficiency of the device, $LHE(\lambda)$, is then given by

$$LHE(\lambda) = 1 - 10^{-OD} \tag{5.2}$$

With an optical density of 1.68, 98% of the light is absorbed. Figure 5.3 shows the electron micrograph of such a nanocrystalline film. When derivatized with *cis*-di(thiocyanato)bis(2,2'-bipyridyl-4,4'-dicarboxylate)-ruthenium(II), the light harvesting capacity becomes superior to that of amorphous silicon. Nature, in fact, uses a similar means of absorption enhancement by stacking the chlorophyll-containing thylakoid membranes of the chloroplasts to form the grana structures.

Figure 5.3:   Scanning electron microscopy of a nanocrystalline titanium dioxide film.

5.2.4   *Light-induced charge separation*

The role of the ruthenium complex is the same as that of chlorophyll in the green leaf: it must absorb the incident sunlight and exploit the light energy to induce a vectorial electron transfer reaction.

In place of the biological lipid membrane, a TiO_2 film is employed. Apart from acting as a support for the sensitizer, it also functions as an electron acceptor and an electronic conductor. The electrons injected by the sensitizer travel across the nanocrystalline film to the conducting glass support serving as a current collector. The driving force necessary for the rapid vectorial charge displacement is small. It corresponds to about 0.1 eV required for the electron injection process to occur irreversibly and on a rapid time scale.

The efficiency of light-induced charge separation at the oxide semiconductor surface is expressed by the quantum yield of charge injection (φ_{inj}). This presents the fraction of the absorbed photons converted into conduction band electrons. Charge injection from the electronically excited sensitizer into the semiconductor is in competition with other radiative or radiationless deactivation channels. Taking the sum of the rate constants of these non-productive channels together as k_{eff} results in

$$(\varphi_{inj}) = k_{inj}/(k_{eff} + k_{inj}) \tag{5.3}$$

One should remain aware that the deactivation of the electronically excited state of the sensitizer is generally very rapid. Typical k_{eff} values lie in the range of $10^6 - 10^9 \, s^{-1}$. To achieve a good quantum yield, the rate constant for charge injection should be at least 100 times higher than k_{eff}. That means that the injection rate has to exceed $10^{11} \, s^{-1}$. In actual fact, in recent years sensitizers have been developed that satisfy these requirements. These dyes should incorporate functional groups ("interlocking groups"), for example, carboxylates, phosphonates and other chelating groups, which, besides bonding to the titanium dioxide surface, also effect, as illustrated in Fig. 5.4, an enhanced electronic coupling of the sensitizer with the conduction band of the oxide semiconductor.

Very promising results have so far been obtained with ruthenium complexes where at least one of the ligands was 4,4'-dicarboxy-2,2'-bipyridyl. The carboxylates serve to attach the Ru complex to the surface of the oxide and to establish good electronic coupling between the π^* orbital of the electronically excited complex and the 3d wave function manifold of the TiO_2 film, see Fig. 5.5.

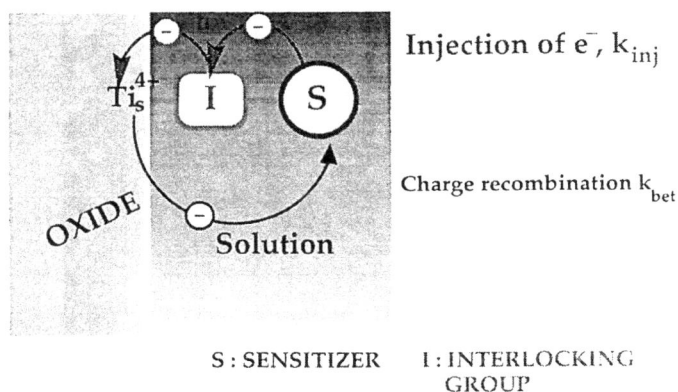

S : SENSITIZER   I : INTERLOCKING
GROUP

Figure 5.4:   Photo-induced charge separation on the surface of titanium dioxide; k_{inj} and k_b represent the rate constants for electron injection and recombination, respectively.

The substitution of the bipyridyl with the carboxylate groups lowers also the energy of the π^* orbital of the ligand. Since the electronic transition is of metal-to-ligand charge transfer (MLCT) character, this serves to channel the excitation energy into the correct ligand, that is, the one from which electron injection into the semiconductor takes place. With molecules like these, the charge injection is in the picosecond or even femtosecond time domain [4].

As the last step of the conversion of light into electrical current, a complete charge separation must be achieved. On thermodynamic grounds, the preferred process for the electron injected into the conduction band of the oxide film is the recapture by the sensitizer cation. Naturally, this reaction is undesirable since instead of electrical current it merely generates heat. For the characterization of the recombination rate, an important kinetic parameter is the rate constant k_b. It is of great interest to develop sensitizer systems for which the value of k_{inj} is high and that of k_b low. Fortunately, for the transition metal complexes we use, the ratio of injection over recapture rate k_{inj}/k_b is greater than 10^3 and in some cases, it exceeds even one million. The reason for this behavior is that the molecular orbitals involved in the back reaction overlap less favorably with the wave function of the conduction

PHOTO INDUCED HETEROGENEOUS
ELECTRON TRANSFER CYCLE

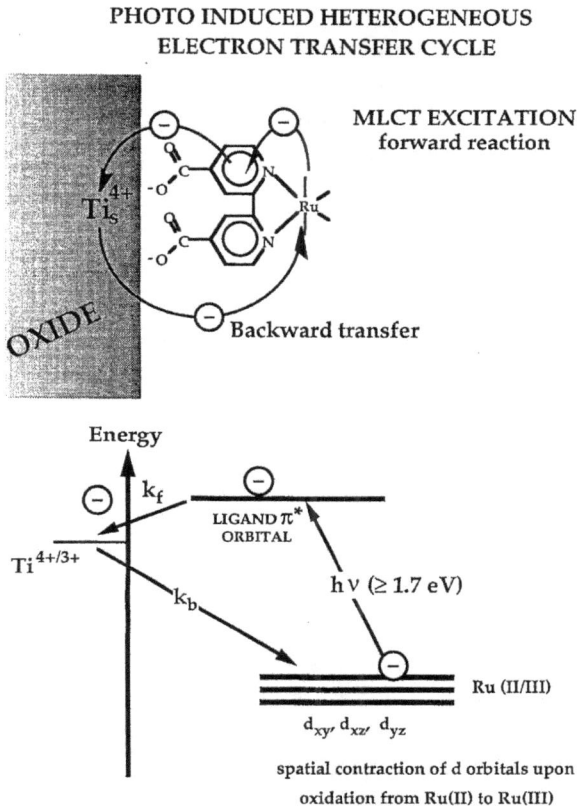

Figure 5.5: Molecular orbital diagram for ruthenium complexes anchored to the TiO$_2$ surface by a carboxylated bipyridyl ligand. The visible light absorption of these types of complexes is a metal-to-ligand charge transfer (MLCT) transition. The carboxylate groups are directly coordinated to the surface titanium ions producing intimate electronic contact between the sensitizer and the semiconductor.

band electron than those involved in the forward process. For example, for *cis*-dithiocyanatobis(2,2'-bipyridyl-4,4'-dicarboxylate)-ruthenium(II) bound to the nanocrystalline oxide film, the injecting orbital is the π* wave function of the carboxylated bipyridyl ligand since the excited state of this sensitizer has a metal-to-ligand charge transfer character. The carboxylate groups interact directly with the surface Ti(IV) ions, resulting in good electronic coupling of the π* wave function with the 3d orbital manifold of the conduction band of the TiO_2. As a result, electron injection from the excited sensitizer into the semiconductor membrane is an extremely rapid process occurring in less than a picosecond, see Fig. 5.5.

By contrast, the back reaction of the electrons with the oxidized ruthenium complex involves a d-orbital localized on the ruthenium metal whose electronic overlap with the TiO_2 conduction band is small. The spatial contraction of the wave function upon oxidation of the Ru(II) to the Ru(III) state further reduces this electronic coupling. This, together with the fact that the driving force for the back electron transfer is large enough to place it in the inverted Marcus domain, explains the relatively low backward electron transfer, which, typically, is in the microsecond time domain.

Thus, in analogy to natural photosynthesis, light-induced charge separation is achieved on kinetic grounds, the forward electron transfer being orders of magnitude faster than the back reaction. As a consequence, the presence of a local electrostatic field is not required to achieve good efficiencies for the process. This distinguishes nanocrystalline devices from conventional photovoltaic cells [5–9] in that the successful operation of the latter is contingent upon the presence of a potential gradient within the p–n junction.

5.2.5   *Photovoltaic performance of transition metal sensitizers*

In our nanocrystalline cell, the electrons injected into the semiconductor from the excited transition metal complex are collected as an electric current, resulting in photovoltaic conversion of light energy. The incident monochromatic photon-to-current conversion efficiency or "external quantum yield" of such a device is given by the equation

$$\eta_i(\lambda) = LHE(\lambda) \times \varphi_{inj} \times \eta_e \qquad (5.4)$$

where $\eta_i(\lambda)$ expresses the ratio of the measured electric current to the incident photon flux for a given wavelength while η_e is the charge collection efficiency. Using ruthenium or osmium complexes of the type shown in Fig. 5.1 in conjunction with nanocrystalline TiO_2 films, solar cells are now available for which all three factors

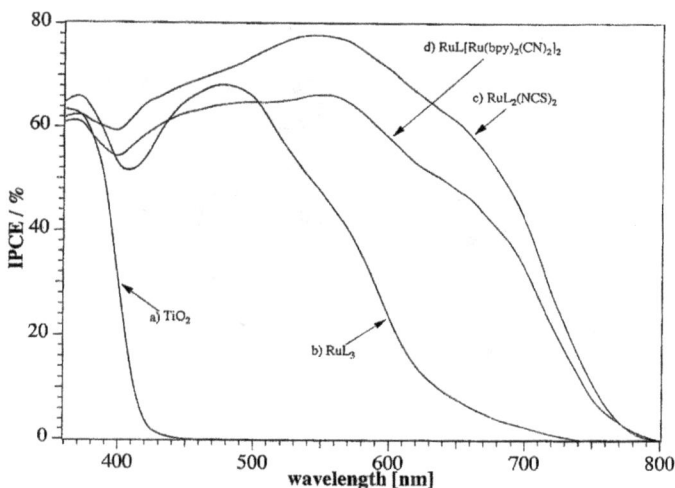

Figure 5.6: Photocurrent action spectrum obtained with three different ruthenium-based sensitizers attached to the nanocrystalline TiO$_2$ film. The blank spectrum obtained with the bare TiO$_2$ surface is shown for comparison. The incident photon-to-current conversion efficiency is plotted as a function of the wavelength of the exciting light.

in Equation (5.4) are close to unity [2, 10–14]. Thereupon, within the wavelength range of the absorption band of the complex, a quantitative conversion of incident photons to electrons is obtained.

A graph which presents the monochromatic current output as a function of the wavelength of the incident light is known as a "photocurrent action spectrum", and Fig. 5.6 shows such action spectra for three ruthenium complexes, illustrating the very high efficiency of current generation, exceeding 75%, obtained with these complexes. When corrected for the inevitable reflection and absorption losses in the conducting glass serving to support the nanocrystalline film, yields of practically 100 percent are obtained. Historically, RuL$_3$ (L = 2,2'-bipyridyl-4,4'-dicarboxylate) was the first efficient and stable charge-transfer sensitizer to be used in conjunction with high-surface area TiO$_2$ films [10]. In a long-term experiment carried out during 1988, it sustained 9 months of intense illumination without loss of performance. However, the

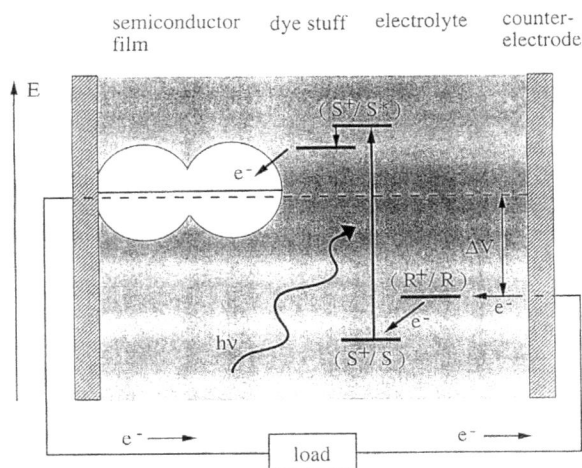

Figure 5.7:   Schematic representation of the principle of the new photovoltaic cell to indicate the electron energy level in the different phases. The cell voltage observed under illumination corresponds to the difference in the quasi-Fermi level of the TiO$_2$ under illumination and the electrochemical potential of the electrolyte. The latter is equal to the Nernst potential of the redox couple (R/R$^-$) used to mediate charge transfer between the electrodes.

visible light absorption of this sensitizer is insufficient for solar light conversion. A large improvement of the light harvesting was achieved with the trimeric complex of ruthenium [1, 2, 11, 16] whose two peripheral ruthenium moieties were designed to serve as antennas [16]. However, the most successful charge-transfer sensitizer investigated so far is, without any doubt, *cis*-dithiocyanatobis(2,2'-bipyridyl-4,4'-dicarboxylate)-ruthenium(II). This achieves close to quantitative photon-to-electron conversion over the whole visible range [2]. Even at 700 nm, current generation is still 40–50% efficient.

The photovoltage of our cell represents the difference between the Fermi level of titanium dioxide under illumination and the redox potential of the electrolyte, see Fig. 5.7. Using the triiodide/iodide redox couple in *N*-methyloxazolidinone solution, under full sunlight an open-circuit cell voltage of 0.7–0.9 V can be measured. Under a 1000-fold lower intensity, the cell voltage is about 200 mV lower — a relative change of cell voltage of only 20–30%. For the conventional silicon cell, the cell voltage decreases by a factor of 3 for a comparable change of light intensity, showing that the photovoltage of our cells is significantly less sensitive to light intensity variations than that in conventional photovoltaic devices. This is an important advantage for the application of the nanocrystalline cell in consumer electronic devices.

A further advantage of the nanocrystalline solar cell is that increasing the temperature from 20 to 60°C has practically no effect on its conversion efficiency. In contrast, conventional silicon cells exhibit a significant decline over the same temperature range amounting to ca. 20%. Since the temperature of a solar cell will reach about 60°C under full sunlight, this feature of our device is particularly attractive for power generation under natural conditions. At this stage, industrial companies are already involved in the production of such cells, which will reach the market quite soon.

The overall efficiency (η_{global}) of the photovoltaic cell can easily be calculated from the integral photocurrent density (i_{ph}), the open-circuit photovoltage (V_{oc}), the fill factor of the cell (ff) and the intensity of the incident light (I_s).

$$(\eta_{global}) = i_{ph} \times V_{oc} \times ff/I_s \qquad (5.5)$$

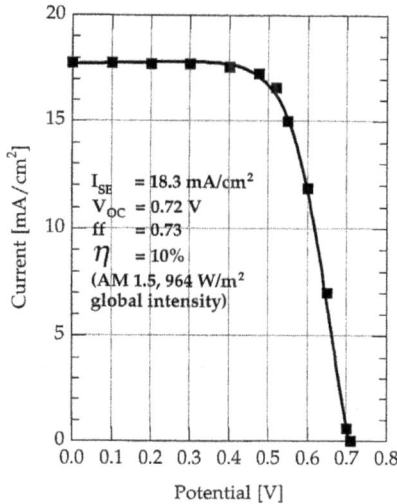

Figure 5.8: Photocurrent–voltage curve for a nanocrystalline photovoltaic cell based on *cis*-Ru(2,2,-bipy-4,4'-dicarboxylate)(SCN)$_2$ as a sensitizer.

In Fig. 5.8, the current–voltage characteristic of a nanocrystalline cell sensitized by I, the photocurrent measured at $96.4\,\mathrm{mW/cm^2}$ of simulated solar intensity was $18.3\,\mathrm{mA/cm^2}$, the open circuit voltage was $0.72\,\mathrm{V}$, and the fill factor was 0.73, yielding for the overall efficiency of the cell a value of 10%.

5.2.6 *Practical development of nanocrystalline solar cells*

Meanwhile, the development and testing of the first cell module for practical applications has begun. The layout of the module is presented in Fig. 5.9. The cell consists of two glass plates, which are coated with a transparent conducting oxide (TCO) layer. The nanocrystalline titanium dioxide film deposited on one plate functions as a light trap. Visible light is absorbed by a monomolecular layer of the ruthenium complex acting as a charge-transfer sensitizer. On illumination, this injects an electron into the titanium dioxide conduction band. The electrons pass over the collector electrode into the external circuit where they perform work. They are then returned to the cell via the counter electrode. The sensitizer film is separated from the counter electrode by the electrolyte. In the electrolyte, there is a redox couple, for example, triiodide/iodide, whose role is to transport electrons from the counter electrode to the sensitizer layer, which had been left positively charged as a result of the electron injection. The counter electrode is made of glass covered with a transparent conducting oxide (TCO) layer that serves as a current collector. A small amount of platinum (5–$10\ \mu\mathrm{g/cm^2}$) is deposited onto the TCO to catalyze the cathodic reduction of triiodide to iodide.

By developing a new mode of Pt deposition, we have engineered an extremely active electrocatalyst, attaining exchange current densities of more than $0.1\ \mathrm{A/cm^2}$ at very low Pt loading. Furthermore, this electrocatalyst is very stable and does not show long-term anodic corrosion as was observed to be the case for Pt deposits produced by conventional sputtering or galvanic methods. These favorable characteristics allowed to minimize overvoltage losses increasing the efficiency and stability of cell operation.

Magnification

(a) 1

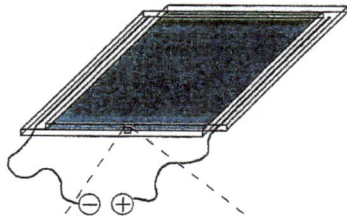

(b) 10^3

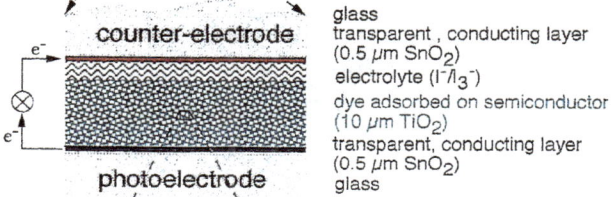

counter-electrode

glass
transparent , conducting layer
(0.5 μm SnO_2)
electrolyte (I^-/I_3^-)
dye adsorbed on semiconductor
(10 μm TiO_2)
transparent, conducting layer
(0.5 μm SnO_2)
glass

photoelectrode

(c) 10^6

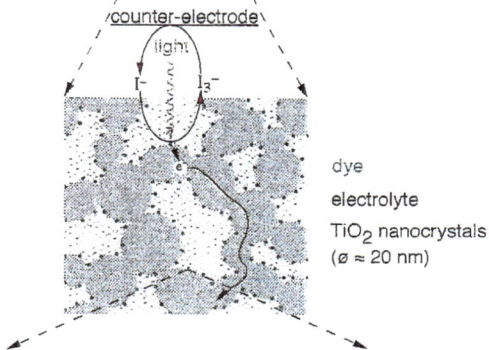

dye

electrolyte

TiO_2 nanocrystals
(ø ≈ 20 nm)

(d) 10^8

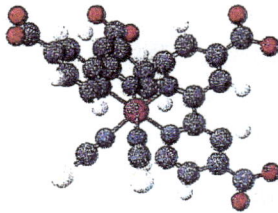

    carbon
    nitrogen
    oxygen
    sulfur
    hydrogen
    ruthenium

Figure 5.9: Layout of a nanocrystalline solar cell. The transparent dye-derivatized TiO_2 film is sandwiched between two conducting layer glass electrodes. It harvests light from both the front and back sides.

Upon long-time illumination, complex I sustained more than 5×10^7 redox cycles without a noticeable loss of performance corresponding to ca. 10 years of continuous operation in natural sunlight. By contrast, practically all organic sensitizers tested so far underwent photo-bleaching after less than 10^4 redox cycles. This clearly outlines the exceptionally stable operation of the newly developed ruthenium charge-transfer sensitizers, which is of great advantage for the practical application of these devices. It may be argued that the presence of ruthenium renders the price of the sensitizer too high for commercial exploitation or that there is insufficient supply of it. However, the required amount of ruthenium is only 10^{-3} moles/m^2 of ruthenium complex corresponding to an investment of ca. US\$0.07/m^2 for the noble metal. One ton of ruthenium alone incorporated in the charge-transfer sensitizer I could provide 1 GW of electric power under full sunlight. This is more than twice the total photovoltaic capacity presently installed worldwide.

Apart from efficiency and stability any future photovoltaic technology will be valued according to its environmental and human compatibility. There is great concern about the adverse environmental effects and acute toxicity of CdTe or CuInSe$_2$, which are being considered for practical development as thin solar cells. Such concerns are unjustified for our nanocrystalline device. Titanium dioxide is a harmless environmentally friendly material, remarkable for its very high stability. It occurs in nature as ilmenite, and is used in quantity as a white pigment and as an additive in toothpaste. Similarly, ruthenium has been used without adverse health effects as an additive for bone implants.

Commercial applications will begin in the near future. Contracts for the industrialization of these cells have been signed with the Wissenschaftspark, Gelsenkirchen for modules of 100 peak watts and higher, which is hoped will lead to applications in the utility market. Meanwhile, for consumer applications cooperation has been established with the Swiss companies SMH, Leclanché, Glass Trösch, Solterra and Solaronix and the first product — a self-powered bathroom scale will be marketed in the course of the present year. In full production, a cost estimate commissioned from the Research

Triangle Institute (North Carolina, USA) predicts a module cost of 60 US cents per peak watt rating.

Quite aside from its intrinsic merits as a photovoltaic device, the sensitized nanocrystalline photovoltaic device will undoubtedly promote the acceptance of alternative energy technologies, not least by setting new standards of convenience and economy for the photovoltaic industry as a whole.

5.2.7 *References*

[1] (a) A. Juris, V. Balzani, F. Barigiletti, S. Campagna, P. Belzer, and A. v. Zelewski, *Coord. Chem. Rev.* 84 (1988) 85. (b) K. Kalyanasundaram, *Photochemistry of Polypyridine and Porphyrine Complexes*, Academic Press, London, 1992, and references cited therein.

[2] M. K. Nazeeruddin, A. Kay, J. Rodicio, R. Humphrey-Baker, E. Müller, P. Liska, N. Vlachopoulos, and M. Grätzel, *J. Am Chem Soc.* 115 1993, 6382.

[3] H. Ritter, "Kythnos photovoltaic power plant", in *Solar Energy R&D in the European Community*, Series C, Volume 1, *Photovoltaic Power Generation: Proc. of Final Design Review Meeting on EC Photovoltaic Pilot Projects*, held in Brussels, 30 Nov–2 Dec, 1981 ed. W. Palz. (R. Reidel Publishing Company, Dordrecht, Holland) p. 217.

[4] R. Eichberger and F. Willig, *Chem. Phys.* 141 1990, 159.

[5] G. Hodes, I. D. J. Howell, and L. M. Peter, *J. Electrochem. Soc.* 139 1992, 3136.

[6] A. Hagfeldt, U. Björksten, and S. Lindquist, *Sol. Energy Mater. Sol. Cells* 27 1992, 293.

[7] D. Liu and P.V. Kamat, *J. Phys. Chem.* 97 1993, 10769.

[8] A Hagfeldt, S. Lindquist, and M. Grätzel, *Sol. Energy Mat. Sol. Cells*, 32 1993, 245.

[9] B. O'Regan, J. Moser, M. Anderson, and M. Grätzel, *J. Phys. Chem.* 94 1990, 8720.

[10] N. Vlachopoulos, P. Liska, J. Augustynski, and M. Grätzel, *J. Am. Chem. Soc.* 110 1988, 1216.

[11] M. K. Nazeeruddin, P. Liska, J. Moser, N. Vlachopoulos, and M. Grätzel, *Helv. Chim. Acta* 73 1990, 1788.

[12] R. Knödler, J. Sopka, F. Harbach, and H. W. Grünling, *Sol. Energy Mater. Sol. Cells* 30 1993, 277–281.

[13] H.J. Lowalt, "300 kW photovoltaic pilot plant Pellworm", in *Solar Energy R&D in the European Community*, ... (*Op. cit.*) p. 179.

[14] T. A. Helmer, C. A. Bignozzi, and G. J. Meyer, *J. Phys. Chem.* 97 1993, 11987.

[15] J. Desilvestro, M. Grätzel, L. Kavan, J. Moser, and J. Augustynski, *J. Am. Chem. Soc.* 107 1985, 2988.

[16] G. Smestad, *Sol. Energy Mat. Sol. Cells* 32 1994, 259.

5.3   Discussion following Michael Grätzel's presentation

Balberg: How do the properties of the cell depend on the concentration of your particles? On the one hand you would want to have as few particles as possible in order to minimize the series resistance. But, on the other hand, the larger the number of particles the more reactions there are that will occur on account of the increased surface area.

Grätzel: We observe a decline in efficiency if the film thickness exceeds 20 μm.

Boxman: You said that these films are now in commercial production. Are there any cost figures and production details that you can share with us?

Grätzel: The first module, manufactured by Leclanché, is used to power bathroom scales. The manufacturers charge about 3 SFr (ca. $2.25) for 20 cm^2, which is about half the cost of a Si solar cell but this cost merely represents today's market forces. The true limiting cost of these cells is the cost of conducting glass. We actually made some comparative measurements on a reference amorphous Si cell manufactured by Darlan (formerly DST). We measured the action spectrum of this grey, almost black, cell and found that the injection efficiency was only 25% at 700 nm. It then increased to a maximum of 75% followed by a decline towards the blue end of the spectrum. Our cell, in contrast, extends further into the red and does not show a blue decline. It consequently gives a larger current. This was the principal reason that it was preferred to amorphous Si by the bathroom scale manufacturer. Another advantage we have over silicon cells is the relatively low-technology production process: No high vacuum is required.

Medwed: Do you currently have any large-scale, such as grid-connected, projects underway?

Grätzel: We have a 1 kW village project with the Chinese Academy of Sciences. This will be the first "large" scale application. But the glass manufacturers will probably create their own large-scale projects in the form of "Smart" windows for industrial and commercial buildings. By using an infrared sensitive dye, we can create a totally transparent (i.e., to visible light) window that will act as a 10% efficient solar cell!

Balberg: Can you say a few words about Frank Willig's experiments?

Grätzel: He has performed laser-injection and charge transport measurements. He finds that the electrons seem to move through the film without noticing that there are centers they can combine with. There could be recombination with the oxidized iodide but this is evidently very slow, as shown by the dark current. The latter is flat and then rises as a typical diode, characteristic of a good rectifying junction. The oxidized dye actually persists for about 10 ns but the electrons simply ignore it.

Yogev: Did you choose your electrolyte because of its refractive index?

Grätzel: No. Its stability properties were far more important. One does not want a volatile electrolyte that evaporates while the cell is in operation. Molten salts are stable in the temperature range from about $-50°$ C to about $300°$ C. That is why they make very good systems. The refractive index of our electrolyte, being $n = 1.5$, is indeed good. The particles themselves have $n = 2.5$ but their 50% porosity leads, via the Bruggeman theory, to a value of $n = 2$ for the film. But the conducting glass support SnO_2 also has $n = 2$, which is why you see hardly any reflection.

5.4 Big dishes for concentrating photovoltaics, solar thermal and other applications (Professor Stephen Kaneff)

Keynote presentation by Stephen Kaneff, Energy Research Centre, The Australian National University, Canberra, ACT, Australia 0200.

5.4.1 *Abstract*

Notwithstanding considerable effort over several decades, electricity generation costs attainable from non-concentrating photovoltaic systems are still unable to compete with those of traditional energy sources, except in a valuable (but relatively small) niche market supplying small to very small units and for certain remote area applications where traditional fossil fuel and related costs are high. However, with the advent of suitable cells and the development of means for effectively cooling those cells, advantage can be taken of higher conversion efficiencies resulting from increased photon flux derived from solar concentrating devices, thereby allowing the use of smaller surface area of the necessary cells, a factor which can have a major cost advantage, even allowing for the need for concentration and cooling.

With well-designed cells and matching systems, the photovoltaic unit costs play a lesser role in overall system cost, a situation which requires that the concentration and cooling systems be cost-effective. Herein lie various limitations which tend to be alleviated by choosing appropriate concentration ratios and economical design of the concentrators themselves. As with many engineering systems, there results an economy of size of the solar concentrator: further advantage occurs from the splitting of the solar spectrum to provide simultaneous photovoltaic electricity and solar thermal outputs (which can also provide electricity if required), as well as waste heat utilization applicable to produce water desalination and/or other co-generation features. Such systems have significant potential for achieving overall system economic viability in the near term.

Use of paraboloidal dish concentrating collectors to drive both photovoltaic and thermal systems together permits greater demand to be established for the dish concentrator technology, which has also widespread application in driving thermal and thermochemical systems. This demand promotes more rapid and more economical progression of the technology. Various developments and potential and economic promise of concentrating photovoltaic systems are discussed, especially in relation to the use of dish concentrators.

5.4.2 *Introduction*

By employing the newly developing photovoltaic cells of appropriate materials and configurations, able to be driven by and to tolerate (given adequate cooling) solar flux concentrated from several suns to more than 1000 suns, there emerges a wide potential field of applications — currently in its early stages of exploitation. A clear promise is apparent regarding economic advantage of this technology in comparison with non-concentrating photovoltaics. Furthermore, by making use of the non-uniform conversion efficiencies from photon flux to electrons across the solar energy spectrum, additional technological and economic value can accrue from splitting the solar spectrum into components which can be directed to provide different outputs simultaneously.

For example, Fig. 5.10 portrays a concentrating split spectrum system in which photon flux from 600 to 1100 nm is employed to drive photovoltaic cells, the cooling fluid for which is employed to assist in preheating the feed water for a solar thermal system which is driven by flux in the range 300–600 nm (for which the effectiveness of PV conversion is not as great as that in the 600–1100 nm range), together with flux above 1100 nm (for which band gap limitations cause difficulties with PV conversion).

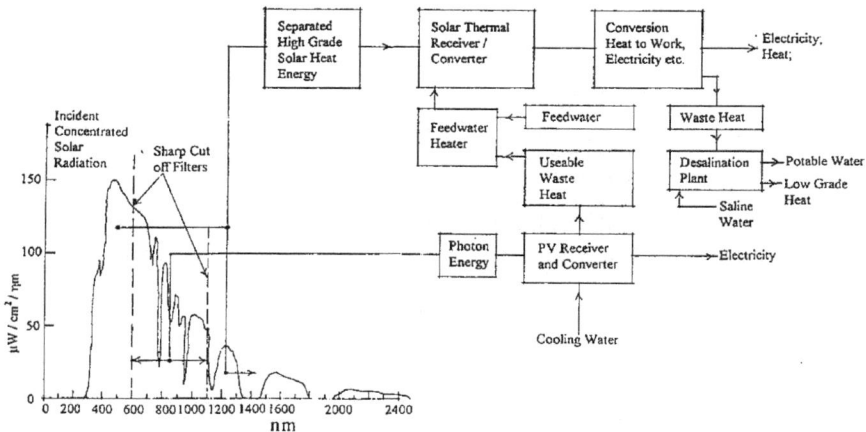

Figure 5.10:   Combined concentrating PV–thermal system.

The solar thermal system itself could provide high quality steam for electricity generation, the waste heat of which could power a water desalination plant (the waste from which in turn could provide lower quality heat for drying or building heating). Other spectrum splitting approaches are practicable.

Hitherto much effort has been applied to the use of parabolic trough PV concentrators (typically of less than 40× suns) and linear and point focus Fresnel lenses. With the advent of large concentrating dishes however, a significant advantage is emerging which allows improved economics over other point or line focus devices. Concentrating dishes have advantages as means for providing enhanced solar flux; they also have advantages in PV/thermal combinations. Other reasons favor dishes as the providers of concentrated solar flux, particularly supplying non-electrical outputs, including applications involving solar-driven chemistry — solar gasification; energy storage by thermo-chemical means; and a wealth of processes dependent on thermal, thermo-chemical and photo-chemical reactions — all solar driven. An economy of size of larger dishes (an economy transferred to dish/PV systems) is now apparent; moreover, due to their multi-purpose functions, dish costs can be reduced by greater demand and the drive for development enhanced. These aspects, together with their already rapidly reducing costs, will ensure that economic viability for applications improves concurrently.

5.4.3   *Rationale for and relevance of concentrating photovoltaics*

As concentration ratio is increased, successively lower investment resides in the photovoltaic cells themselves, while more is required for the concentrating and cooling hardware; the latter therefore tend to determine the overall economics. That is, to an increasing extent as cells develop, the economic viability of concentrating photovoltaic systems depends on the availability or realization of economically viable solar concentrators. Fortunately, such concentrators are becoming well developed and are attaining good economic viability, especially in the case of paraboloidal dish concentrators of high concentration ratio and large size. Consequently, the practicability of concentrating PV systems depends very largely, at present, on the availability of the cells themselves.

Concentrating photovoltaic cells have much reduced area, to an extent determined by the concentration ratio required. While the cells are more expensive than 1 sun PV cells, they turn out to be less expensive per unit of module aperture area. Thus, for a given levelized energy cost, concentrating PV cell costs can be tolerated to rise significantly if concentration ratio is increased, as indicated in Fig. 5.11, which shows relationships between cell cost and cell peak efficiency at different concentration ratios. The advantages of high concentration ratios are apparent, so long as cell cost — and balance of system cost — can be contained.

In the example cited in Fig. 5.11, a target levelized electricity cost of 12 cents/kWh may be achieved within a near term (5 years) in the case of cell costs amounting to one third of total module costs, with effectiveness and concentration ratios within practicable limits. While it is not intended here to recount the continuing development of concentrating PV cells and their efficiencies, it may be noted that such efficiencies in realizable cells are increasing and are tending towards economic reality somewhat as indicated by

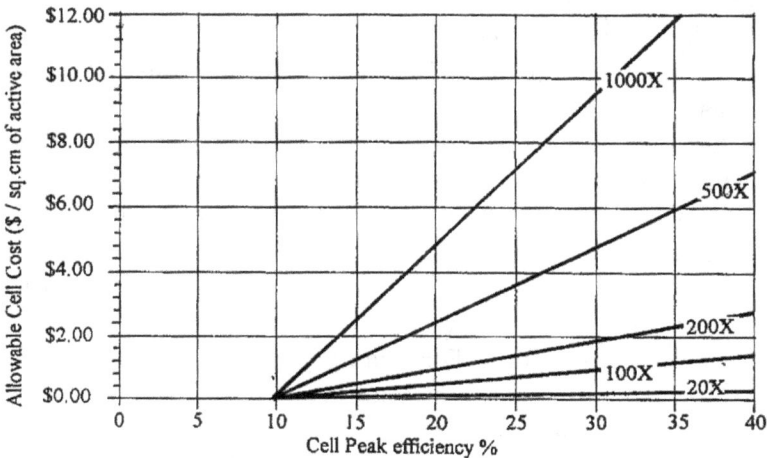

Figure 5.11: Allowable maximum cell costs vs. cell peak efficiency, at different solar concentration ratios, to achieve a levelized energy cost of 12 cents/kWh at a cell cost of 1/3 of the total module cost (from [1]).

the characteristics of Fig. 5.11. Thus a wide range of strategies is becoming available for the realization of concentrating PV systems, for example, utilizing the following:

- low efficiency (15–17%)/low concentration ratio (10–30 suns)/low cost cells ($0.04–0.20/cm^2), employing line focus modules of polycrystalline silicon X-Si;
- high efficiency (24–30%)/medium concentration ratio (100–500 suns)/medium cost cells ($1–4/cm^2) with point focus concentrators and high-efficiency X-Si cells;
- very high efficiency (30%+)/high concentration ratios (500–1000 sun)/higher cost cells ($6–12/cm^2) ensuring point focus concentrators and gallium arsenide (GaAs) multi-junction cell technologies or combined cells of various configurations.

In summary, concentrating PV systems to provide a competitive edge compared with non-concentrating units, depend on the following:

- reduced overall energy cost as a result of increased efficiency;
- the fact that fewer modules are required for a set output;
- less supporting structure being required for the cells;
- less land area required for the installation;
- lower installation cost for the cells (but this needs to be considered in conjunction with the concentrator costs);
- achieving cost effectiveness in the concentration and cooling systems. This aspect is assisted by the potential for co-generation and solar thermal/PV combined systems, as illustrated in Fig. 5.10;
- a suitable accessible market being available for concentrating PV systems. Such a market is cost-driven and can be identified in specific situations. For example, Table 5.1 lists a number of townships in Western Australia where diesel-based power is expensive, and concentrating PV systems which achieve a levelized electricity cost of 12 cents/kWh or lower could be cost-effective now. Overall in Australia, there are many diesel stations producing electricity for remote areas, with a total capacity of some 500 MWe. Many installations are small

Table 5.1: Costs of power in remote grids in Western Australia.

Location	Total Cost c/kWh	Capacity Site Rating kW	Marginal Fuel Cost c/kWh	Peak Demand kW	% Over Peak	Proj Growth % year	% Safe Over Peak
Broome	11.81	13720	8.69	8200	67	4	18
Camballin	22.02	893	17.90	270	230	2	85
Carnarvon*	9.56	16506	5.20	8830	87	2	33
Cue	12.60	1152	12.16	530	117	4	40
Denham	17.49	1465	14.33	760	93	4	0
Derby	15.46	11100	9.11	5300	109	2	41
Esperance	12.43	14406	10.29	10190	41	4	1
Exmouth	13.92	8776	8.75	4300	105	2	44
Fitzroy	15.65	2564	12.10	1230	108	5	37
Gascoyne		240	26.77	60	300	2	100
Halls Creek	15.46	2668	12.19	1130	136	2	56
Hopetoun	16.20	1014	12.00	390	160	3	59
Kununurra	15.56	12400	10.73	6620	87	5	32
Argyle	23.86	570	18.45	130	338	2	100
Leonora	16.16	3389	11.48	1730	95	5	38
Marble bar	24.20	1386	14.77	490	182	1	67
Meekatharra	14.46	3740	11.88	2020	85	3	21
Menzies	23.61	350	17.37	80	357	5	113
Mt. Magnet	13.52	3360	12.47	1770	89	3	−3
Nullagine	24.42	370	14.95	190	94	4	0
Onslow	17.45	2064	13.88	910	126	2	33
Ravensthorpe	16.02	2024	11.65	730	177	4	66
Sandstone	26.85	235	23.29	70	235	3	71
Wiluna	11.48	1014	10.65	340	198	3	82
Wittenoom	23.78	800	16.57	210	280	2	124
Wyndham	17.56	4336	11.04	1850	134	0	28
Yalgoo	19.58	285	17.22	120	137	3	25
Total	18.88	110.832	13.74	58.455	190	32	49

Data from State Electricity Commission of Western Australia 1991, Excluding Pilbara Grid, Norseman and Laverton. * = on gas.

(average 300 kW) and carry fuel costs of 12–20 cents/kWh, with larger systems liable for fuel costs of 9–12 cents/kWh. This situation is mirrored elsewhere in the world. In many regions, no electricity exists and consequently opportunities arise to provide for new installations. The situation is that suitable cost-effective concentrating PV systems do not rely on substantial technological breakthroughs, but do require a commercialization phase. This places them in an arguably better position than that of non-concentrating PV cells, for which a long-standing unmet promise of attaining mainline grid-like costs still seems a long way off, depending as it does on substantial advances. The technology for achieving cost-effective concentrating PV systems, however, is already at hand.

It is relevant to underline the fact that while the economy of size of concentrating collectors works to improve the economics of concentrating PV systems, there is no added system efficiency to be derived from multiple concentrating collector/PV arrays. This is unlike the case of solar thermal systems for which arrays of collectors feeding larger turbines can take advantage of the increase of turbine efficiency (and reduction in cost/MWe of the turbogenerators) as well as allowing the use of efficiency and economic enhancement by using topping and bottoming cycles cost-effectively. However, if a number of concentrating PV collectors are employed, an economy of number of units does arise.

In view of these and other considerations, it appears probable that concentrating PV systems will find their place in supplying electricity competitively in systems initially below 1 MW. (Because of the need for tracking for all but low concentrating systems, very small concentrating PV units have difficulty in competing because of the overheads involved; consequently, there would expectedly be a minimum economic size yet to be determined.) As higher efficiency cells of lower cost evolve, this boundary can be expected to expand upwards. Whether such systems will compete with larger solar thermal systems employing fossil or biomass fuel involving combined cycle systems (for example, gas turbine/steam turbine combinations with solar steam input and later employing the products of solar gasification) remains to be resolved, but this latter technology is also fast developing.

5.4.4 *Means for achieving concentrated photovoltaic and related systems*

Configurations of great ingenuity and often complexity have been reported in attempts to harness concentrating cells and to take advantage of spectrum splitting. The application of line focus parabolic troughs as in Fig. 5.12(a) — and line and point focus Fresnel lens systems as in Fig. 5.12(b) — has already been noted for lower concentration ratios. These approaches represent the more common directions for development, but are not necessarily the most cost-effective.

Other more complex systems have been proposed, for example, spherical light confining cavities as illustrated in Fig. 5.13, whereby sunlight from a concentrator enters and illuminates an arrangement of solar cells. Light reflected from these cells most likely falls on the cells again; thereby reducing the apparent reflectance at the light entry point. An ideal transformer at the entry aperture converts the rays from a primary to an isotropic bundle, and the spherical shape becomes isotopically illuminated by the aperture. These mechanisms reduce the apparent reflectance, increase light confinement and increase the concentration at which the cell can operate [2].

Figure 5.14 illustrates the schematic layout of a "double tandem" arrangement in which three successive cells are "illuminated" and

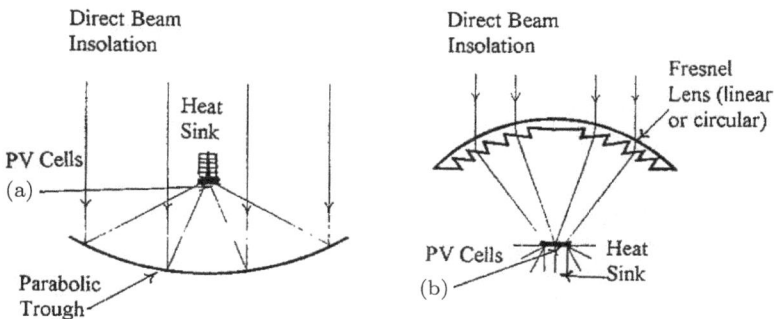

Figure 5.12:   Concentrating PV cells illuminated by (a) parabolic trough system, and (b) Fresnel lens system.

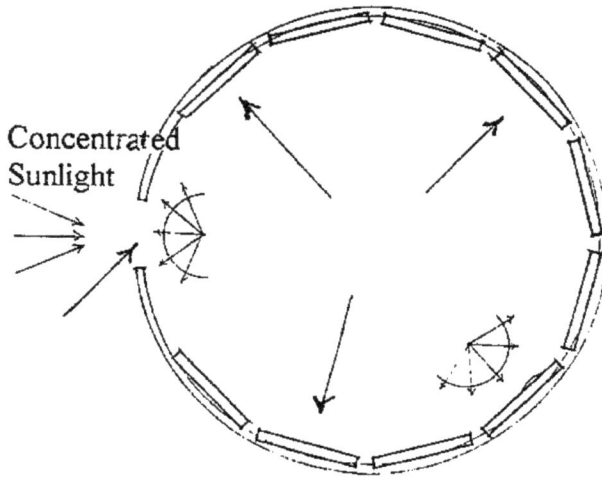

Figure 5.13:   Spherical light confining cavity (from Ref. [2]).

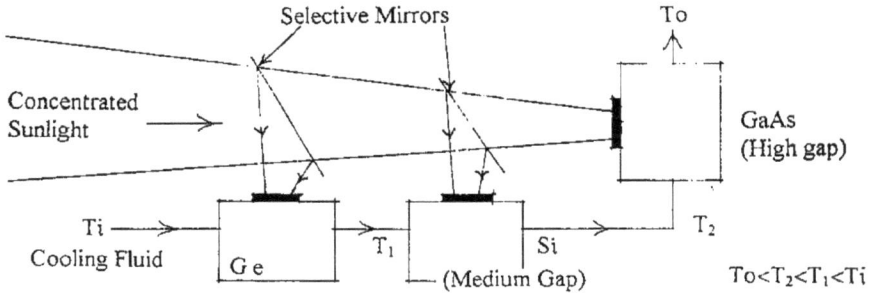

Figure 5.14:   Employing the "Double Tandem" principle for a three-cell configuration, with cooling means provided.

cooled, whereby the temperature of the cooling fluid first cools the
cell with the lowest band gap and then the higher gap cells. Using
high concentration ratios (approx. 1000), theoretical efficiencies of
45–50% are assessed overall — with the heat energy driving a thermal
power engine [3].

A characteristic of much of the reported proposals for concentrating PV systems of high performance all too often is the omission

of experimental achievements without which practical assessment is difficult. Programs of development of high-concentration photovoltaic modules to redress these limitations to advances are underway (e.g., [4]).

5.4.5 *Photovoltaic and solar thermal power —*
a comparison

Environmentally, solar thermal (dish-based) systems are currently at an advantage compared with non-concentrating PV systems in relation to carbon emissions avoided and low energy payback period [5]. Solar thermal/thermochemical systems also have the potential and means to provide much more than electricity (which represents the lesser fraction of our total energy used) and process heat. Solar gasification can give gaseous and liquid fuels and other solar-driven chemical reactions can provide a vast array of energy-rich products and especially energy storage [5] — indeed most of our energy could be supplied via such paths.

Waste heat from solar thermal systems may be used in many co-generation configurations, particularly when employed with combined cycle gas turbine/steam turbine combinations which have no counterpart in PV. Associated conversion efficiencies on solar only can exceed 40% overall and ongoing developments are set to increase this figure when system sizes are large — tens to hundreds of megawatts. The largest solar thermal units so far built — 80 MWe (Luz) — represent only a beginning, as thermochemical systems of 1000 MWe and larger (with storage) can already be envisaged.

Installed and generation costs of solar thermal systems are lower than achieved PV costs and are set to decline further as collector and overall system costs take advantage of latest developments in the field; indeed these developments are still very much in their infancy and solar thermal systems can already be built to compete with mainline power systems in many situations [6].

PV systems, on the other hand, are able to satisfy the need for very small powers extremely conveniently and are reaching effectively towards larger sizes. Concentrating PV appears to have a better potential on the above comparison with solar thermal than non-concentrating PV. But better still, the theme is here stressed that combined concentrating PV/solar thermal combinations should be advantageously pursued, particularly in sizes up to megawatts or so, for which there appears a current advantage.

5.4.6   *Co-operative concentrating photovoltaic solar thermal systems*

While very large solar thermal systems have the advantages of high efficiency and low installation and generation costs, smaller systems relying on lower turbine efficiencies are not so favored. Accordingly, in view of the relatively high conversion/efficiencies achievable from concentrating PV systems and their enhancement by combining solar thermal output as well, these latter systems have the potential for attaining good economic viability in smaller than megawatt sizes, thereby rivaling purely solar thermal units. Starting from this position it is not a major step to extend the sizes of combined concentrating PV–solar thermal systems to much larger units as the economics of collectors and PV cells advance. This represents the best of both worlds. Figures 5.10 and 5.14 are indicative of ways of achieving this co-operation.

The system of Fig. 5.10 has an additional aspect of co-generation to enhance output further by employing the solar thermal "waste heat" to desalinate water, a process which can be most effectively achieved as regards exergetic consumption by employing multiple-effect distillation as indicated in Fig. 5.15, which provides a comparison between various desalination approaches [7].

In some situations, while the electricity produced by both the PV and solar thermal sections is valuable, the purified water can be even more so.

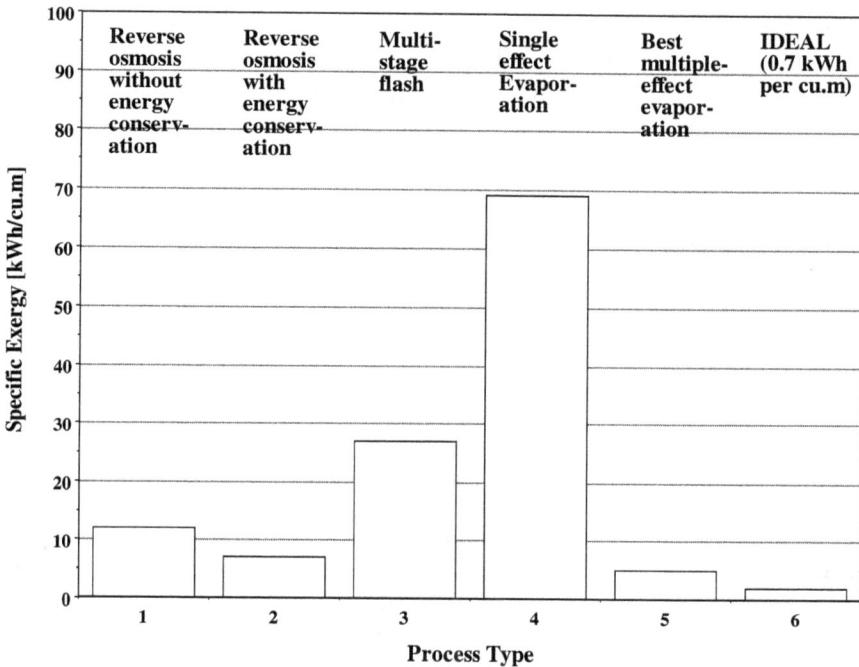

Figure 5.15: Exergetic consumption of various desalination processes.

5.4.7 Advantages of dish-based PV concentration systems

Features include the following [8]:

- Dishes are multi-purpose and can provide economical concentrated photon and heat flux to meet a wide range of flux concentration characteristics;

- The high concentration ratios readily attainable from dishes allows the use of very small receivers which, apart from having low losses, much more importantly require only relatively small areas of concentrating cells. While the small size of receiver presents advantages, cooling systems need careful attention in design and protection;

- Dishes provide the highest conveniently realizable concentration ratios, for which PV cell efficiencies are high;
- Because the PV module represents only a fraction of the overall cost of the system and is physically a small part, changes are readily encouraged in experimental systems, and replacement is practicable in commercial systems;
- With spectrum splitting, both PV and thermal drive are available;
- Dish systems are now well developed, are commercially available; and large dishes present attractive economics;
- Above all, economics of dish concentrating PV and combined concentrating PV/solar thermal (+ co-generation) are appearing to be very attractive.

Dish systems are of comparatively recent development and few reports of successful experimental results are apparent in the literature, although much ongoing study is evident. Lasich *et al.* [9] have reported a successful demonstration with a 1.5-m dish in which a single close-packed silicon array produced more than 200 watts with an efficiency of 22% at 239 suns and a gallium arsenide module produced 85 watts with 18% efficiency at 381 suns. They expect complete receiver efficiencies of over 20% to result. This expectation is matched by that of Blakers and Keough at the Australian National University, who have carried out experiments on in-house-produced silicon cells using an Energy Research Centre 5 meter dish during 1994.

Johnston's work in the characterization of the $400\,\mathrm{m}^2$ dish is moving to the next stage involving the design and fabrication of a tandem photovoltaic/thermal receiver, initially to be tested on a 5 meter diameter dish. Concentrator silicon photovoltaic cells will be placed behind a wavelength selective filter in the focal region, permitting PV cells to convert 600–1100 nm wavelengths at the highest efficiency. Other wavelengths will be absorbed in a thermal receiver to produce heat of 150–200°C.

These and many other structures elsewhere are at an early stage and will provide interesting results over the coming years.

5.4.8 *Illustrative comparative costs*

Non-concentrating PV and Trough- and Dish- based Concentrating PV

Table 5.2 summarizes comparative costs of three PV systems, for sizes of a few hundred square meter collector area in each case (after manufacture facilities exist).

Table 5.2: Summary of electricity costs for various PV systems.

Electricity Generation System	System Cost $/m^2$	System Efficiency %	Levelized Cost of Electricity
PV dish, small scale production ($ 12000 for cells) (1)	390	16–17%	9>/kWh
PV trough (based on production run) (2)	400		15>/kWh
Flat plate module (based on production run) (3)	460		24>kWh

In Table 5.2, relatively low efficiencies are taken for the dish PV system. (1) is estimated from Kaneff [8] for a 400 m^2 aperture dish. (For the first-off unit, a cell cost of \$50,000 is allowed.) Levelized electricity costs are based on 8% real discount rate, 2% O&M, 30 year dish life, and 2360 kWh/m^2/year installation. (2) and (3) are taken from Keogh *et al.* [10]; trough concentration ratio = 14. Due to difficulty in precise comparison, costs are indicative only. With higher efficiency cells, the dish system would be more competitive. No allowance is made for manufacturer's/supplier's profit. Cost estimates are based on rough design only. The comparison suggests, however, the ordering of respective costs.

5.4.9 *Other applications for dish systems*

The nature and form of dishes allow modular construction and expansion as an inherent feature so that systems can be built smaller or larger. Much if not most applications can potentially be provided via dishes, including the following:

- Electricity and process heat generation (preferably distributed dish central plant systems);
- Solar-driven chemistry — solar thermo-chemical systems;
- Solar gasification and production of syngas from coal and biomass;
- Production of liquid fuels;
- Energy storage and the provision of base load solar energy.

5.4.10　*Large aperture dishes — 400 m² for PV systems*

Current dishes produced by the Energy Research Centre have 400 m² aperture and are considered suitable for concentrating PV applications due to their relatively low cost.

Figure. 5.16 portrays the Mark I dish configuration which consists of an extremely light weight space truss system of great rigidity. The 54 triangular mirror surfaces conform to a paraboloidal shape and may be assembled from plane mirror segments or can be curved.

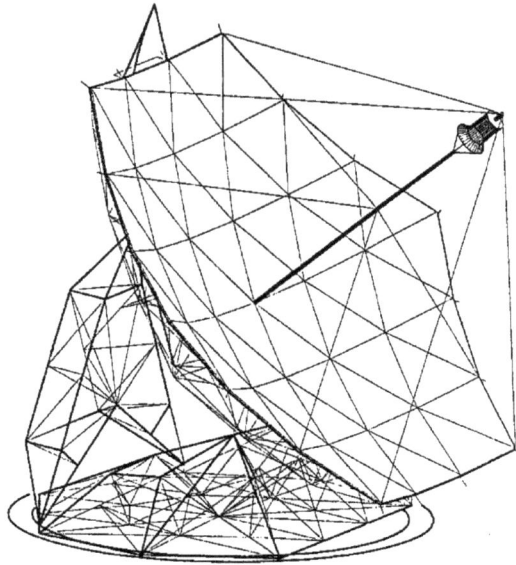

Figure 5.16:　Paraboloidal dish, 400 m² aperture; Mark I.

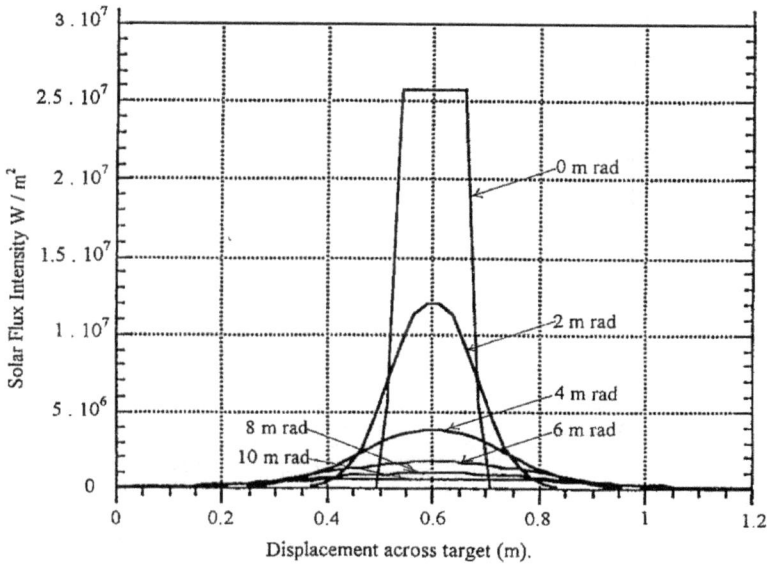

Figure 5.17: Assessed solar flux distributions for 400 m^2 aperture dish at the focal plane (13.1 m) for various dish surface errors.

The dish itself is actuated in azimuth and elevation, to track by hydraulic actuators controlled by an electronic controller to follow the sun irrespective of cloud. Figure 5.17 indicates the assessed effect of various average slope errors of the mirrored surface. In the case of the present dish, a slope error 6 milliradians was designed and achieved; as indicated by Fig. 5.18. This specific design target was required in order to alleviate the flux density distribution on the receiver. The actual focal plane flux intensities are indicated in Fig. 5.19, which shows reasonable characteristics.

A feature of this dish design is the ability to provide desired flux concentration ratios to suite needs. For example, were the 54 triangular panels to hold plane glass mirrors, a concentration of some 54 suns is achieved segmenting each of the 54 panels into a number of flat panels, but held to conform to the paraboloidal surface by piecewise approximation, other concentration ratios can be achieved (9 segments/panel, produce $9 \times 54 = 486$ suns). With curved mirrors high concentration ratios can be provided, expectedly

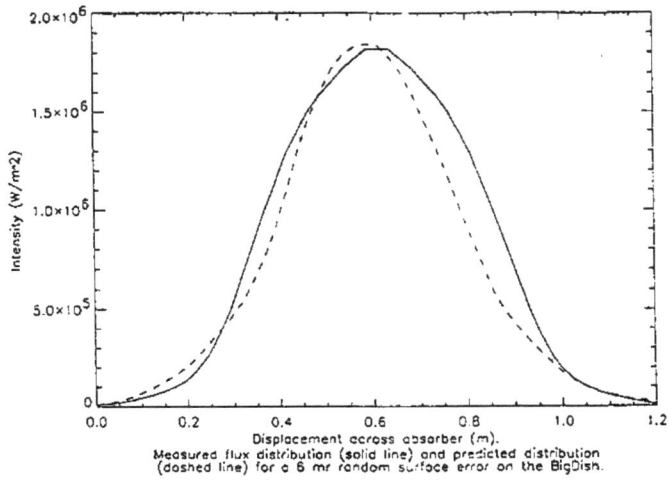

Figure 5.18: Confirmation of average slope error of $400\,\mathrm{m}^2$ aperture dish (6 milliradians as designed).

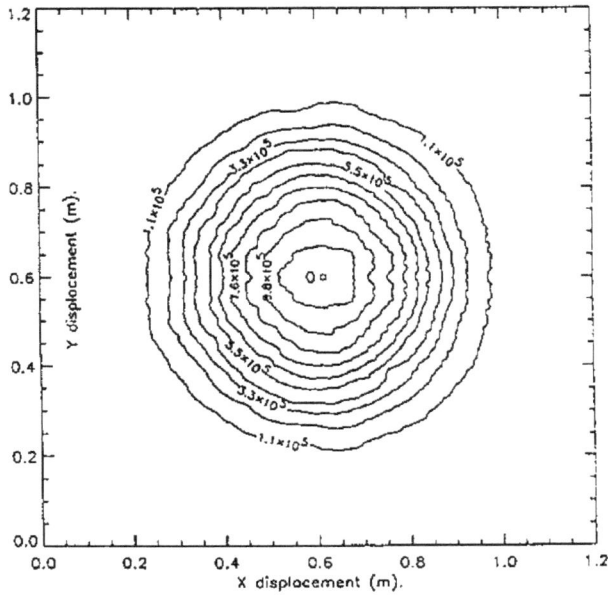

Figure 5.19: Focal characteristics of 400 m^2 aperture dish designed for solar thermal steam generation. Numbers represent direct beam radiation levels in W/m^2 for an incident direct beam radiation of $875\,\mathrm{W/m}^2$.

well above 5000 suns. The accuracy and rigidity of the frame permit these characteristics to be established and maintained.

Finally, by adjusting selected panels, several foci can be provided in close proximity to allow simultaneous systems to operate.

5.4.11 *The future*

Whatever the long-term future holds, in the near term it seems very probable that concentrating PV with dishes will present itself as an economical option to the production of electricity in appropriate situations. As with all forms of renewable energy, this should provide a complementary benign energy source, most probably of increasing viability, especially in combination with solar thermal energy, utilizing dish collectors.

5.4.12 *Acknowledgments*

The generous support to attend this symposium, provided by Professor David Faiman and the symposium organizers, is gratefully acknowledged. The overall field of concentrating collectors and solar power has been long supported by the Australian National University, ANUTECH, the Energy Authority of New South Wales and, for a period, by Allco Steel in relation to paraboloidal collectors. Since early 1993, additional support has come from the Electricity Trust of South Australia, the Federal Energy Research and Development Corporation, the Electricity Supply Association of Australia, Office of Energy of New South Wales, Pacific Power (NSW), Power and Water Authority of the Northern Territory, the Queensland Electricity Commission and the State Electricity Commission of Victoria. All of this assistance is acknowledged. Special thanks go to Glen Johnston who performed the calculations and measurements for Figs. 5.17–5.19.

5.4.13 *References*

[1] A. B. Maish and J. L. Chamberlain, *PV Concentrators Today and Tomorrow*, Sandia National Laboratories, Albuquerque, NM, 1991.

[2] A. Luque, G. Sala, J. C. Miñano, P. Davies, I. Tobias, J. Alonso, C. Algora, J. M. Araujo, A. Cuevas, and J. Olivan, "The photovoltaic eye: a high efficiency converter based on light trapping and spectrum splitting", *Proceedings Tenth European Photovoltaic Solar Conference*, Lisbon, April 1991, p. 627.

[3] A. Goetzberger, W. Bronner, and W. Wettling, "Efficiency of a combined solar concentrator cell and thermal power engine system", *Proceedings of the Tenth European Photovoltaic Solar Energy Conference*, Lisbon, April 1991, pp. 10–14.

[4] W. P. McNaughton, R. D. Cummings, F. J. Dostalek, and R. H. Rickman, "Phased development of high-concentration photovoltaic modules", *Prog. Photovol. Res. Appl.* 1 1993, 107–131.

[5] D. L. Hagen and S. Kaneff, *Application of Solar Thermal Technologies in Reducing Green House Gas Emissions*, Report to the Australian Government Department of the Environment, June 1991, 305 pp.

[6] S. Kaneff, *Mass Utilisation of Solar Energy*, Energy Research Centre Report, EP-RR-63, The Australian National University, 1992.

[7] S. Kaneff, *The Use of Solar Energy for Water Desalination*, Energy Research Centre Report, ER-RR-64a, The Australian National University, 1993.

[8] S. Kaneff, *Solar thermal Process Heat and Electricity Generation: Performance and Costs for the ANU Big Dish Technology*. Energy Research Centre Report, ER-RR-57, February 1991.

[9] J. B. Lasich, A. Cleeve, N. Kailc, G. Ganakas, M. Timmons, R. Venkata-subramanian, T. Colpitts, and J. Hills, "Close packed cell arrays for dish concentrators", *Proceedings of the First World Conference on PV Energy Conversion*, Hawaii, December, 1994, 3 pp.

[10] W. Keogh, A. Blakers, A. Cuevas, and D. Mills, *PV Trough Feasibility Study*, Department of Engineering and Information Technology, The Australian National University, May 1994, 56 pp.

5.5 Discussion following Steve Kaneft's presentation

Boxman: How does your kind of dish compare in cost with the mass-produced satellite dishes one sees on roof tops?

Kaneff: The target price of our dish — under conditions of mass production — is less than AUD 100,000 (ca. US$75,000). This works out at less than US$190/m^2; I do not know the cost of satellite-tracking antennas but it is probably not lower than this.

Roy: Where does your glass come from and what kind is it?

Kaneff: Our first dish was made from locally available glass of 2 mm thickness that was given to us by Pilkington. We subsequently purchased some 1 mm glass from a Swiss company at AUD $30/m^2$, ready silvered and cut to size.

Biryukov: To what extent is high reflectance important for your dishes?

Kaneff: It is extremely important, as is the need to keep the surfaces clean. At our White Cliffs power station, we employ manual cleaning using lamb's wool pads but the 400 m^2 dish, in power system applications, would employ mechanical cleaning.

Koltun: How can you improve the uniformity of illumination in your target plane?

Kaneff: With our particular dish design, the focal area can be made more uniform in one of two ways. One can use flat mirror panels in order to produce a naturally smeared out focal spot. Alternatively, if the mirror panels are paraboloidal, one can offset them so that each produces its focal spot in a slightly different location.

Gordon: By degrading the focal spot to produce a more uniform illumination of the target, you throw away much of your collected energy. Would it not be more sensible to use secondary and perhaps tertiary optics to achieve uniform illumination with minimal loss of energy?

Kaneff: Your criticism about wasting energy is correct only if the smearing process is carried out grossly by adjusting the 54 large panels. However, by adjusting the many flat segments that constitute each panel, one could, in practice, achieve a fairly uniform focal region with little loss. Your suggestion about using secondary optics is well taken; however, it requires much ingenuity to realize these ideas. I know of one particular attempt that ended up requiring an enormous cooling system to keep the second-stage concentrator from overheating. So if you don't lose energy one way, you are likely to do so in another way.

Yogev: If I would buy one of your dishes tomorrow, what would be the slope error, the tracking error and the cost?

Kaneff: The dish we have working in Canberra has a measured average slope error of 6 milliradians: deliberately large since we don't want to melt the steam boiler at the focus. The tracking error is considerably better: probably about 1 milliradian but we have not yet verified this figure. As for the cost, it all depends upon what precisely you want to buy. Only the dish? And if so, an exact copy of the Canberra dish or one with a smaller mean slope error? I've already stated what our target price should be. We are currently running at a factor of about 3 higher for a one-off item.

Yogev: I also have a comment. In my opinion, the accuracy figures you quote are not good enough for commercially viable systems. Thermal power-producing engines reach their maximum efficiency in the multi-megawatt size and for such engines, your dishes are not large enough. That leaves concentrator photovoltaics. But for this to be cost-effective, you need ultra-high concentration together with spectrum splitting for efficient use of all parts of the radiation.

Kaneff: Permit me to disagree with both of your points. Based on many years of engineering experience I believe that dish thermal systems of a few megawatts will be commercially viable and as many dishes as required can be used. This was the motivation that led us from the small White Cliff dishes to the present $400\,m^2$ model. Our philosophy was to start small and keep things as simple as possible. I believe I could make a dish that would be as accurate as you like if all that needed to be satisfied were intellectual curiosity. But this is not a commercial motivation. Concentrator photovoltaics naturally has its own requirements. If the PV modules prove themselves to be cheap and reliable, then our dish can be made to have whatever degree of sophistication is appropriate. I would point out, however, that concentration ratios greater than 1000 suns, although easily achievable by the dish, are unnecessary for concentrator photovoltaics as presently envisaged.

Grätzel: Do you think concentrator photovoltaic systems will stand up to the thermal stress of high concentration for tens of years?

Kaneff: I don't think that anyone can answer that question yet. I would only point out that people were similarly skeptical about our

White Cliffs solar-thermal project. Yet the system worked for 14 years without any materials problems and I am sure that it could continue working for another 14 years.

Balberg: One of your slides indicated that concentrator photovoltaics would be more cost-effective than non-concentrating modules. How general is that result supposed to be and why isn't everyone using concentrator modules?

Kaneff: Concentrator modules are not yet commercially available. But the result you mention referred to a calculation for a specific site under rather specific economic conditions. It is not supposed to be universally true. For example, you could not use concentrator systems in a cloudy location.

Chapter 6

1996

6.1 Editor's Foreword

The *7th Sede Boqer Symposium on Solar Electricity Production* was almost entirely devoted to photovoltaics. Our two, highly celebrated, keynote speakers were Keith Barnham of Imperial College, London, UK and Keith Emery of NREL, Boulder, CO, USA. However, that symposium was also graced by one of the titans in the history of photovoltaics: Namely, Morton Prince, who, living part of the year in Jerusalem, was a frequent attendee at these symposia.

Dr. Prince was one of the original co-developers at Bell Labs of the world's first practical PV cells. He went on to develop cells for use on the Vanguard artificial satellites, the first of which was launched in 1958. President Jimmy Carter subsequently appointed him director of the country's first solar energy R&D program, which later became known as the US Department of Energy. This editor vividly recalls one of Morton's visits to Sede Boqer. He asked how our (unsuccessful!) attempts were going at producing an all-carbon pn junction cell using the newly discovered C_{60}. He had read about the Schottky-barrier cell we had succeeded in making, and that its efficiency was only about one thousandth of one percent due to the fact that we were forced to use intrinsic C_{60}. When I asked what kind of efficiencies he and his colleagues had obtained with intrinsic silicon back in his Bell Lab days, he exclaimed: "Intrinsic silicon? We used industrial silicon and had no idea what kind of impurities it contained!"

Keith Emery was famous for the novel and uniquely accurate PV characterization methods he developed in his laboratory at NREL, for which he received several prestigious awards. For many years, the *only* believable claims to world record cells were those that had been certified by NREL! Another personal recollection was when this editor received a telephone call from Keith sometime after his keynote lecture at Sede Boqer. He was aware from our published papers that Sede Boqer possessed a *natural* AM1.5 global sky spectrum (under suitable geometric and atmospheric conditions). At that time, there was renewed interest in concentrator PV cells. However, the original computer program that had generated both the well-known AM1.5G standard and lesser-known but defective AM1.5D spectra was no longer readable, and the notes for constructing the program had been lost. Keith asked whether we could provide him with a pair of *natural* AM1.5 global and direct spectra to help NREL reconstruct a new AM1.5D standard.

Professor Barnham established one of the world's leading photovoltaic research laboratories, at Imperial College, London. Within that framework, he pioneered the growth of quantum wells in photovoltaic cells: devices which, as he explains in his keynote lecture, improve upon both the current and the voltage compared with standard solar cells of the same material. Throughout his long career he has continued to play a leading role in the development and commercialization of such cells. Again on a personal note: Although we never met one another at the time, Keith's early work as an experimental physicist in the area of elementary particles confirmed one of this editor's theoretical predictions in that field. So it was mutually satisfying for both of us finally to meet up in our newly and independently chosen field of solar energy.

6.2   Photovoltaics past and present:
A decade-by-decade review of photovoltaics
(Dr. Morton Prince)

An after-dinner talk given by Morton B. Prince, Jerusalem

As you know from the program, this slot was supposed to be occupied by Professor Luque who, at the last moment, was prevented from arriving, by force major. The symposium organizers asked me whether I might be prepared to improvise something and, as luck would have it, I happen to have with me a set of slides I prepared last year as part of a talk [1] I gave at the 12th European Photovoltaic Solar Energy Conference. [Editor's note: The "talk" was an acceptance speech upon being awarded the 1994 Becquerel Prize. The interested reader may care to look up W.H. Bloss's laudatory speech [2] in honor of Dr. Prince, in the Proceedings of that conference.]

When one looks at the progress that we have obtained in photovoltaics on a day-to-day or week-to-week basis, it may seem that we are not even moving. But if one looks at the progress on a decade-by-decade basis, we have made fabulous advances. There is an English language saying — "we cannot see the forest for the trees" — and this applies to photovoltaics as well as to most other human endeavours.

So now let us look at the last four decades and try to project only the next decade. It is only by looking at this macro-picture that we see just how rapidly this technology is moving.

In Fig. 6.1, I show a time-line with the various decades that I will review very briefly up to the present, with representative names for each decade and I am going to mention what I believe are the highlights in these decades in the six categories: Materials; Techniques; Devices; Applications; Information-Transfer; and Commercial Operations.

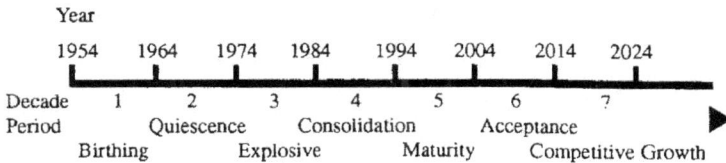

Figure 6.1: Time-line.

Figure 6.2 describes the first decade which I have named the *Birthing Period*. The major highlight of this decade was the announcement by the Bell Telephone Laboratories of the silicon solar battery in April 1954. Although some research was carried out on

1. **Materials:**
   Single crystal silicon, GaAs, CdS/Cu$_2$S.
2. **Techniques:**
   Czochralski crystal pulling, diffusion, ion implantation (1963), dendritic growth (1963), sputtering.
3. **Devices**:
   Si p-n junction, linear arrays (Si) for punched-card readouts, spherical. Si ball array cells, thin Si cells (space).
4. **Applications**:
   Space power, consumer products (radios, flashlights, toys), emergency call boxes.
5. **Information Transfer**:
   1st Symposium devoted entirely to PV — Los Angeles, December 18-19, 1959 — sponsored jointly by AIAA and IRE (37 attendees).
   IAPG (1st IEEE PVSC), April 14, 1961.
   PVSC 2 and PVSC 3.
6. **Commercial Operations**:
   In the U.S. — Hoffman Electronics, International Rectifier Corporation, Spectrolab.
   Outside the U.S. — none.

Figure 6.2: Decade 1 (1954–1963) *Birthing Period*: Major Highlight — Bell Telephone Announcement of Solar Battery, April 1954.

GaAs and CdS/Cu$_2$S devices, almost all the work was carried out on crystalline silicon. In this decade, almost all the cells were made for space applications including the development of thin (100 micron) cells. Also I presented a paper on cells made up of small spheres of silicon. Attempts were made to commercialize terrestrial applications but most of these were unsuccessful since the users were wary of these new devices and they were relatively expensive. This pointed out the need for extensive education of the general public in the value of photovoltaics. The first meeting devoted entirely to photovoltaics was that held in December 1959 on establishing methods and standards for measurements especially for space. At that time there were only 37 attendees. In the United States, the Interagency Power Group held periodic meetings on various power developments and sponsored the first IEEE PVSC in April 1961. During this decade, the only commercial manufacturers were in the United States although the USSR produced some cells for their space program and I saw some of these at the World Fair in Brussels in 1958.

> 1. **Materials**:
>   Single crystal Si, cast silicon (1973).
> 2. **Techniques**:
>   Silicon casting, passivation, thin-film CdS/Cu_2S, EFG ribbon.
> 3. **Devices**:
>   Diffused Si, radiation hardened lithium doped Si.
> 4. **Applications**:
>   Still primarily space, initial terrestrial experimental systems.
> 5. **Information Transfer**:
>   IEEE PVSC #4 thru #10.
> 6. **Commercial Operations**:
>   New U.S. companies — COMSAT, Solarex, Solar Power.
>   European companies — RTC. Ferranti.
>   Japanese companies — Sharp.

Figure 6.3:   Decade 2 (1964–1973) *Quiescence Period*: Major Highlight — Cherry Hill, New Jersey Meeting — October 1973.

The next decade. summarized in Fig. 6.3, which I have named the *Quiescence Period*, since very little was done in this decade for terrestrial purposes, had its major highlight at the end of the decade; it was the famous Cheny Hill meeting in which the attendees let their imaginations run wild to develop a plan for an extensive growth of terrestrial use of photovoltaics. This meeting occurred as an outgrowth of the oil embargo created by OPEC. In the earlier part of the decade almost all the activities focused on space applications including the commercialization in Europe and Japan. In the United States, at the end of the decade, two commercial firms started operations specifically for terrestrial devices: Solarex and Solar Power.

The third decade, the *Explosive Growth Period*, was a period of rapidly growing governmental support. It is depicted in Fig 6.4. The major highlight was the United States federal photovoltaic program which grew to over $150,000,000 in fiscal year 1981. Many strong technical teams were organized which worked on a variety of new photovoltaic materials and devices. Applications were fielded in a variety of grid and non-grid connected tests and terrestrial consumer products using amorphous silicon were exploited by the Japanese manufacturers. The first European Community PVSEC

1. **Materials**:
   Amorphous materials, lll-V (various), organics, CulnSe$_2$, CdTe, Zn$_3$P$_2$, Cu$_2$O, etc.
2. **Techniques**:
   Ribbon growth (several), glow discharge, CVD (MOCVD), electrodeposition, homoepitaxy, heteroepitaxy, laser processing, spray pyrolysis.
3. **Devices**:
   Heterojunctions, TPV, PEC, MIS, BSF, Fresnel concentrators, fluorescent concentrators, grain boundary devices, optical confinement, bifacial, inverters, power conditioning components.
4. **Applications**:
   Residential, commercial and industrial stand-alone and grid connected applications; consumer products (calculators, watches, etc.); utility uses; microwave repeaters; etc.
5. **Information Transfer**:
   IEEE PVSC's
   EC PV Solar Energy Conference — 1st held in September 1977
   PV Generators in Space — 1st Symposium held in September 1978
   SPRAT Conferences in the US
6. **Commercial Operations**:
   In the U.S. - Sensor Technology, Shell, Motorola, Mobil, M-7, Solec International, UPG, Varian, Westinghouse, Astropower, Boeing, Crystal Systems, Glasstech, Chronar, Alpha Solarco, Omnion, Abacus, etc.
   In Europe — Siemens, BP, Photowatt International, Italsolar, Ansaldo, Helios, Nukem, AEG Telefunken, IT Power.
   In Japan — Sanyo, Hitachi, Fuji Electric, Mitsubishi, Kyocera, NTT, Showa Shell, Matsushita.
   Elsewhere — India (CEL), Brazil (Heliodynamica).

Figure 6.4:   Decade 3 (1974–1983) *Explosive Period*: Major Highlight — Government Programs, ERDA/DOE, EEC, Japan.

was held and space conferences were held in Europe and in the United States. A very large number of new manufacturers entered the field throughout the entire world. This decade was truly the flowering of photovoltaics.

The fourth decade, shown in Fig. 6.5, which recently ended has seen a *consolidation* of photovoltaics in preparation for rapid future growth. Many new materials, techniques and devices were developed and applications extended to village power in developing countries and non-central station utility uses. This latter set of applications has

---

1. **Materials**:
   Ternary and quaternary Ill-V's, II-VI's, and I-III-VI's.
   Porous silicon (1993).
2. **Techniques**:
   CLEFT, novel beam techniques for depositing controlled films.
3. **Devices**:
   Multi-junction amorphous materials, spherical Si cells, improved structures of silicon devices.
4. **Applications**:
   Village power, water pumping, utility non-central station.
5. **Information Transfer**:
   IEEE PVSC's,
   EC PVSEC's,
   1st PV Science and Engineering Conference (November 13-16, 1984).
6. **Commercial Operations**:
   In the U.S. — Sunpower, Amonix, Texas Instruments, EPV.
   In Europe — Solems, Isophoton, RES.
   In Japan — Hoxan, Kaneka, etc.
   Elsewhere — India (several), China, Taiwan, etc.

Figure 6.5: Decade 4 (1984–1993) *Consolidation Period*: Major Highlight — Expansion of PV Applications to Utilities.

exposed the utilities to the value of photovoltaics and a large number of them have initiated photovoltaic use in their region throughout the world. During this decade, in addition to the meetings continuing to be held in the United States (IEEE) and in Europe (EUPVSEC), the Pacific Rim started having its series of meetings. Large utility systems of 1 MW or larger continued to be installed.

Now it is time to attempt to project the future: a perilous task! However, I am willing to stick out my neck for the next decade. I call this decade which we are entering the *Maturity Period* (Fig. 6.6), since I believe that much of what we have already developed is now being incorporated into production lines and that many potential users now understand the value of photovoltaics including many of the international financial organizations that can help the developing countries. In fact, that is where I foresee the explosive growth in this decade. In addition we are continuing to develop newer materials, techniques and devices for the more distant future to enable us to get the cost of photovoltaics down to the level needed for utility peaking and even base load needs.

1. **Materials**:
   Ultra-stable encapsulants, theoretically-designed new materials.
2. **Techniques**:
   High volume beam techniques for depositing large area devices with controlled compositions.
3. **Devices**:
   Multi-junction polycrystalline thin film devices, devices using engineered materials, highly-efficient light trapping structures, concentrator subsystems using very high-efficiency multi-junction III-V materials.
4. **Applications**:
   Advanced utility non-central station applications, rapid acceptance in third world for village power and other electrical needs.
5. **Information Transfer**:
   Combined meetings separated by longer time periods, electronic mail permits rapid transfer of information, publication and intellectual rights will have to be redefined with the new communication techniques.
6. **Commercial Operations**:
   Will spread widely throughout the world as demand grows.

Figure 6.6: Decade 5 (1994–2004) *Maturity Period* Major Highlight — Explosive Growth of PV in the Third World.

6.2.1 *References*

[1] B.P. Morton, *Proceedings of the 12th European Photovoltaic Solar Energy Conference*, Amsterdam, The Netherlands 11–15 April, 1994, eds. R. Hill, W. Palz and P. Helm, H.S. Stephens & Associates, Felmersham, UK, Vol. 1, 1994, pp. xl–xliii.
[2] W.H. Bloss, *Proceedings of the 12th European Photovoltaic Solar Energy Conference*, *Op.Cit.*, Vol. 1, pp. xxxvii–xxxix.

6.3   Photovoltaic materials toward the year 2000: Their characteristics and State of the Art (Eng. Keith A. Emery)

A keynote lecture presented by Keith A. Emery (National Renewable Energy Laboratory, Golden, CO, USA). Please note that although this invited keynote lecture was presented by Mr. Emery, the paper that he provided for publication in the Symposium Proceedings — which is reproduced here — was co-authored by Lawrence L.

Kazmerski (National Renewable Energy Laboratory, Golden, CO, USA).

6.3.1 *Abstract*

The status and directions of the photovoltaic technologies are presented. Current and projected research and development directions are indicated. Device technologies are discussed, stressing both research and laboratory progress, as well as commercial- and industrial-level cells and modules.

6.3.2 *Introduction*

Photovoltaics has experienced remarkable research and development progress over the past 10 years [1–4]. The technology is now evolving, during which it is expected to undergo rapid transformations from the research and pilot line phases to real, cost-effective deployment into the user marketplace [5]. The challenges for those working in this energy-producing technology are decision making, identifying and making ready viable markets, ensuring adequate funding, transferring technologies from the laboratory to the production line, enhancing the production and manufacturing capabilities of an embryonic industry, and, most important, building consumer, government, and user confidence in photovoltaics.

The purpose of this paper is to review and update the status and development pathways of the current photovoltaic technologies, with emphasis given to the US Program, the mission of which is to develop photovoltaics for the large-scale generation of economically competitive electric power and making photovoltaics a significant part of the US and world energy mix. The program's revised goals are summarized in Table 6.1 in terms of module efficiencies, electricity price, system lifetime, and US sales for the mid-term (2000) and long-term (2010–2030) periods [5, 6].

A significant part of this program is the aggressive development of an industrial base required for photovoltaics to penetrate various energy end-use sectors. The program includes efforts to form partnerships among manufacturers, utilities and other users,

Table 6.1:   Summary of US department of energy photovoltaics program goals.

Factor	Current Milestones (1995)	Current Plan Goals (2000)	Long-Term Goals (2010–2030)
Module Efficiency (%)	7–17	10–20	15–25
Electricity Price ($/kWh, 1995$)	0.25–0.40	0.12–0.20	0.05–0.06
System Lifetime (years)	10–15	20	30
U.S. Cumulative Sales (MW)	175	500–700	>10000

universities, and federal and state agencies. This paper presents the status of various cell and module technologies, and provides a summary of some of the manufacturing and government activities. Primarily, those cell and module efficiencies that have been measured in the NREL laboratories (under standard test conditions) are cited as confirmed [7, 8]; other reported cell/module performances are discussed.

Thermophotovoltaic (TPV) energy converters produce power by photovoltaic conversion of photons from a radiant heat source. Potential TPV applications include remote electrical supplies, hybrid vehicles, co-generation, electric-grid independent appliances, satellite, aerospace and military applications. There have been two conferences summarizing the state-of-the-art of TPV [4].

6.3.3   *Photovoltaics industry*

Worldwide, the photovoltaics industry remains a dedicated, yet relatively small part of the total energy production enterprise. Until recently, photovoltaic manufacturers and related industry partners have been largely dependent on consumer products and government/political decisions for a significant portion of their livelihood. The industry has fought very hard to expand markets and bring credibility for this renewable resource in the worldwide power-generation arena. International attention has shifted gradually, but significantly, from the concerns of the oil crisis in the late 1970s, to some indifference in the early 1980s, to the climate and environmental issues in recent times. International markets for photovoltaic power

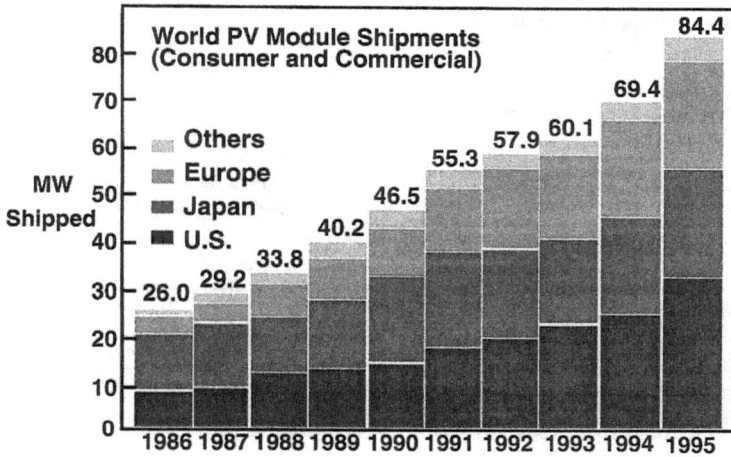

Figure 6.7: Terrestrial photovoltaic module shipments over the time period 1986 through 1995, excluding modules made from purchased cells, but including modules made from produced cells, and excluding photovoltaics for space applications.

are now apparently expanding, and gradual growth for the industry can be anticipated. Three centers for photovoltaic production currently dominate: Europe, Japan, and the United States. The world PV module shipments are depicted graphically in Fig. 6.7, covering the time period 1986–1994 [9]. These have more than doubled over this time frame, and more than more than 70 MW were shipped in 1994. Dominating these shipments are the single-crystal and polycrystalline Si technologies, accounting for more than 70% of the total in 1991, 1992 and 1993 [9]. The general trend of decreasing product price with increasing production is shown in Fig. 6.8 [9].

Amorphous Si has contributed about 25% of the total shipments, but the majority of these have been consumer products (watches, calculators, etc.) rather than terrestrial power generating resources. Other technologies, especially the very promising thin-film CdTe and $CuInSe_2$ approaches, are still developing in the industry sector, and must start to penetrate the markets in the near future to ensure their viability. The focus for these technologies is certainly to bring them from the laboratory to the commercial phases as their manufacturing,

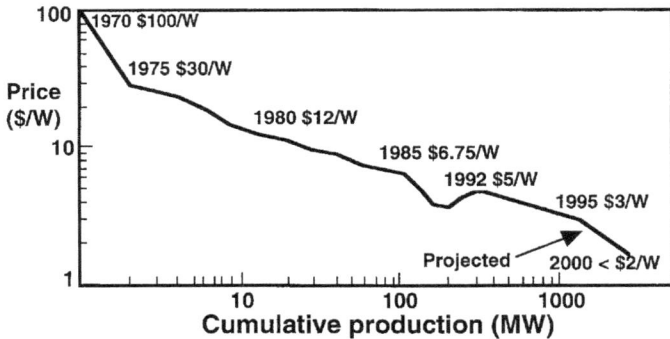

Figure 6.8:   Module experience curve, showing decrease in price as function of production level.

cost, and reliability are demonstrated, and as the industry (and user) shows its commitment and confidence with these newer photovoltaic approaches.

The space photovoltaics industry has continued along at a steady level with satellites using mostly crystal high efficiency silicon cells. Recently there has been an increased demand for cells with higher end of life efficiencies than silicon can achieve [10]. This has led to the increased production of GaAs cells on GaAs or preferably Ge substrates. Multijunction GaInP/GaAs based cells are currently being manufactured at the pilot line level. Currently the estimated annual production requirements for communication satellites is at the 400 kW/year level. The proposed Teledesic communication system will require about 5.5 MW of PV power by the year 2001 [10]. This represents a substantial increase in production of space qualified PV cells over current levels.

The thermophotovoltaics industry is in it's infancy. The markets are just being identified and the research is in its infancy [4]. There have been several prototype systems for military applications and remote electrical sources. It is expected that there will be a rapidly increasing TPV industry as markets are developed and demonstration systems are deployed. TPV appears especially promising as a substitute for thermoelectric based power sources. There are also potential military applications where a minimum of moving parts is essential. The utility industry is interested in self powered gas furnaces where TPV would supply the power for the electric blowers and control system [4].

6.3.4 *Silicon and III–V technologies*

These solar cells are characterized by very high efficiency performance, but at relatively high cost. These technologies are primarily based upon single-crystal semiconductor approaches (although some multicrystalline and thick film approaches are included), and encompass cells that are planned for concentrator, as well as flat-plate applications.

6.3.4.1 *Silicon*

Silicon has been the foundation for the photovoltaics industry from its inception. Silicon solar cells have demonstrated lifetimes, having performed adequately both in space and terrestrial environments. The major limitations have been in the related areas of cost and performance. Much of the current advanced research activity is directed to bringing materials and processing costs down and improving the performance of devices toward predicted values. Ongoing research continues to provide understanding on the role of defects and impurities in altering the properties of the material and the Si cells. In addition, the cell design has undergone rapid evolution. Current device engineering has optimized cell configurations to minimize both photon and minority-carrier losses. Several single-crystal cell designs have been verified with efficiencies exceeding 20%. These designs have a minimized metal semiconductor contact area for the front grid and rear substrate contacts. They typically employ fine-line photolithography to minimize sheet resistance losses, and multiple layer antireflection coatings and in some cases special texturing to minimize reflection losses. The highest confirmed Si cell efficiency under standard reporting conditions at one-sun has a passivated emitter and rear locally diffused back (PERL) from the University of New South Wales (UNSW) [8, 11]. Modules using these advanced cells have recently been reported above 20% efficiency, for both flat-plate and concentrator types [12–14].

Interest in less energy-intensive processes, which sacrifice the crystalline perfection and perhaps use less pure starting materials, remains for polycrystalline and other innovative approaches. The trade-off is usually the ultimate performance or efficiency of the devices. Bulk and ribbon approaches for providing "sheet" Si have undergone extensive investigations; and casting, directional solidification, heat exchanger method (HEM), dendritic web, and edge-fed film growth (EFG) have received commercial attention. Recent improvements in device engineering (following those developed for single-crystal cells) have improved the performance of these bulk polycrystalline Si cells to the 17–18% level (Table 6.2) [8, 15]. Production cells using cast material are still typically in the 11–12% range.

The Photovoltaic Manufacturing Technology (PVMaT) project sponsored by the US Department of Energy and administered by NREL has resulted in 30% reductions in module cost with improved performance in Siemens czochralski Si modules through lower wafer cost, higher yields, reduced labor in cell and module fabrication, and reduced cost of module materials [16]. The stated goal of Solarex's PVMaT cast polysilicon program is to reduce the module cost to less than US$1.20/W through casting of larger ingots, reduced kerf loss from the wire saws, transfer of higher efficiency cell processes to the manufacturing, increased module assembly automation, development of a frameless module, and automated handling of large thin wafers [17].

Relatively thin-layer cells, grown by a modified liquid phase growth of the Si onto conductive ceramic substrates, have been reported by Astro Power [18]. The process has excellent promise for low-cost production of the cells, and efficiencies exceeding 15% have already been confirmed on small area devices [8]. This technology has reached the pilot plant stage, and a 20 kW array was installed at the PVUSA site in Davis California [19]. Another innovative approach using cells fabricated on Si-spheres has been announced by Texas Instruments, Inc., termed the Spheral^{TM} Solar Cell Technology [20]. However, this technology is currently "on hold" because of change in technology emphasis at TI, and loss of some concurrent investment.

Table 6.2: Summary of confirmed efficiencies for silicon-based technologies.

	Data	Area (cm²)	Voc (mV)	Jsc (mA/cm²)	FF (%)	Efficiency (%)	Structure/Comments
one-sun Cells							
UNSW	9/94	4.00	709	40.9	82.7	**24.0**	PERL, crystalline [11]
UNSW	4/93	45.7	694	39.4	78.1	**21.6**	PERL, crystalline
Georgia Tech.	12/95	1.00	636	36.5	80.4	**18.6**	HEM, multicrystalline
Sharp	3/93	100	610	36.4	77.7	**17.2**	textured, crystalline
AstroPower	12/88	1.02	600	31.4	79.2	**14.9**	Silicon-Film™
AstroPower	3/95	240	582	27.4	76.5	**12.2**	Silicon-Film™
Aust.Nat.Univ.	9/94	4.02	651	32.6	90.3	**17.0**	20µm thick, thinned Si
Modules							
Honda/SunPower	2/94	862	32.6V	0.703A	81.3	**21.6**	48 crystalline cells
Sandia	10/94	1017	14.6V	1.36A	78.6	**15.3**	HEM, multicrystalline
Texas Inst.	9/94	3931	20.1V	0.692A	73.6	**10.3**	Spheral™ Si
concentrator Cells							
Stanford Univ.	5/87	0.150				**26.5**	140 suns, point contact
Sun Power	7/93	1.21				**25.7**	74 suns,rear contact
UNSW	9/90	20.0				**21.6**	11 suns, laser-grooved
Concentrator Modules							
Sandia/UNSW/Entech	4/89	1875				**20.3**	80 suns, 12 cells module

Notes: The independently confirmed efficiencies in this table are from [8] or from the authors laboratory at NREL. All measurements were performed under standard reporting conditions of 25°C, 1000 W/m², ASTM E892 global spectrum. The area definition used for non-concentrator cells is total area. For concentrator cells the area is the designated illumination area (total area minus peripheral bus bars). The reference spectrum for concentrators is the ASTM E891 direct normal reference spectrum and its one-sun is arbitrarily taken to be 1000 W/m². The aperture area (total area minus frame area) is used for modules.

Quebec hydro is currently developing this process. The spheres are formed by melting millimeter-sized particles of electronic-grade material. The particles tend to take round shapes during melting due to the dominance of surface tension over other forces. As the crystals solidify, single-crystal spheres are formed, and the impurities tend to accumulate near the surface of the sphere. Mechanical grinding then removes this portion of the impurities. This melting step is repeated several times to bring the spheres to the desired purity. Solid state treatments, such as O precipitation and P gettering, are used to passivate the remaining impurities within the sphere. The 0.75-mm diameter spheres then undergo a P-diffusion to create the junction. The cell spheres are processed into an embossed Al foil in an hexagonal pattern to provide a flexible matrix to hold them in place, to contact the spheres electrically, to mask the backside for etching, and to provide a back reflector for the light. Cells with an average efficiency above 10% have been fabricated, with yields of about 90% for 10 cm^2 areas of the flexible foil (about 50 spherical solar cells). Prototype module efficiencies in the 8–9% range have been achieved [21].

6.3.4.2 *III–V cells*

Promise and demonstration of the highest efficiencies have come from III–V single-crystal semiconductors. Table 6.3 summarizes both single-junction and multiple-junction approaches using primarily GaAs and InP. Noteworthy in the single-junction cases are the NREL 25.3% GaInP(window)/GaAs homojunction (0.250 cm^2 area) [22], the Kopin 25.1% GaAlAs (window)/GaAs cell (3.91 cm^2 area) [23], and the Spire 16 cm^2, 24.2% GaAs cell. In the multiple-junction arena, five technologies have been verified with efficiencies exceeding 30%. The first is the Boeing GaSb/GaAs four-terminal, mechanically stacked cell, which was measured at 34% under 100× concentration [24]. The second is a GaAs cell stacked on Si, measured by Sandia National Laboratories. The next three are of NREL design and realization: a three-terminal, monolithic InP/GaInAs tandem cell measured at 31.8% (100 suns) and a four-terminal, mechanically stacked GaAs/GaInAsP cell measured at 30.2% (40 suns) [25]. The highest efficiency III–V monolithic two-junction device is the NREL GaInP/GaAs cell, with an efficiency of 29.5% (AM1.5) and 30.2% (180 suns concentration) [26]. Table 6.3 provides a summary of these devices.

Module technologies using III–V cells have advanced. One approach using thin, single-crystal cells separated from a reusable substrate, the CLEFT (cleaved lateral films for transfer) process developed by Kopin Corp., has reached $4 \, cm^2$ areas with 23.3% efficiency [23]. A CLEFT cell (16 such devices) mini-module with 21% efficiency has been produced. Boeing has recently reported a 25% GaAs module. The major consideration of these technologies is cost. Certainly they have immediate consideration for high-value applications such as space power. However, the cost demands of the terrestrial markets are still impeding the acceptance of these material approaches in competition with alternatives by many photovoltaic decision makers.

6.3.5 *Thin films*

6.3.5.1 *Amorphous technologies*

In the past 15 years, amorphous Si:H research cells have improved steadily from about 2.5% to greater than 12% [2]. Some 10 US and more than 15 Japanese and European groups have reported cells with >10% efficiency. Table 6.4 presents a summary of the better cells with independent confirmation. The industry goals for the amorphous program are for a 10% efficient commercial thin-film module by 1995. It is planned to achieve these stable prototype amorphous modules through a program aimed at better understanding and improving the electro-optical properties of the a-Si:H-based alloy materials. Recently, the US Program has undergone a change in emphasis: from single-junction approaches tolerating *initial efficiencies* to multijunction modules research and development using *stabilized efficiencies* as benchmarks of performance [5, 6]. Stability is still the key issue with this technology, and progress toward understanding light-induced degradation and minimizing its effects remains a significant part of the program.

Table 6.3: Summary of confirmed cell efficiencies for III–V based technologies.

	Date	Area (cm²)	Voc (mV)	Jsc (mA/cm²)	FF (%)	Efficiency (%)	Structure/Comments
one-sun cells							
ASEC	3/89	4.003	1035	27.57	85.3	**24.3**	GaAs, GaAlAs window
Kopin	11/89	4.000	1011	27.55	83.8	**23.3**	GaAs, Cleft (separated)
	3/90	3.910	1022	28.17	87.1	**25.1**	GaAs, GaAlAs window
	4/90	16.00	4034	6.55	79.6	**21.0**	5-mm Cleft GaAs, submodule
SMU	7/85	1.009	822	19.70	62.2	**10.1**	Thin-film GaAs/Ge/graphite
Spire	3/89	0.250	1029	27.89	86.4	**24.8**	GaAs, GaAlAs window
	3/89	0.250	1018	27.56	84.7	**23.8**	GaAs(MBE) Purdue
	4/90	4.015	878	29.29	85.4	**21.9**	InP
	11/88	0.250	1190	23.80	84.9	**24.1**	GaAs/Ge Tandem
	4/92	16.14	1035	26.9	85.4	**24.2**	GaAs (large-area cell)
Varian	3/89	0.500	2403	13.96	83.4	**27.6**	Two-terminal tandem
		0.500	1402	13.92	86.8	**16.7**	1.93-eV GaAlAs top cell
		0.531	1000	13.78	83.0	**11.3**	GaAs bottom cell
		—	—	—	—	**27.3**	Three-terminal tandem (total)
NREL	3/89	4.000	1045	27.60	84.5	**24.4**	GaAs, GaAlAs window
	8/90	0.310	876	28.70	82.9	**20.9**	InP top cell
		0.312	337	21.94	72.1	**5.3**	0.75-eV GaInAs bottom cell
		—	—	—	—	**26.2**	InP/GaInAs 3-terminal tandem
	6/93	0.250	2385	13.99	88.5	**29.5**	GaInP/GaAs 3-terminal tandem
	12/89	0.173	1038	28.70	86.4	**25.7**	GaInP/GaAs, with aperture
	6/93	0.250	1049	28.50	84.8	**25.3**	GaAs heteroface, GaInP window
	8/19/88	0.108	813	27.97	82.9	**18.9**	ITO/InP

Table 6.3: (*Continued*)

	Date	Area (cm^2)	V$_{OC}$ (mV)	J$_{SC}$ (mA/cm^2)	FF (%)	Efficiency (%)	Structure/ Comments
concentrator cells							
NREL	1/91	0.0746	959	1509	87.3	**24.3**	InP homojunction 52 suns
	2/91	0.0746	899	6343	82.5	**27.5**	1.15-eV GaInAsP, 171 suns
	12/94	0.103	2663	2320	86.9	**30.2**	GaInP/GaAs 2-terminal, 178 suns
	10/90	0.0511	1096	990.3	83.5	**23.1**	GaAs top cell, 39.5 suns
		0.0534	626	556.7	80.7	**7.1**	0.9-eV GaInAsP under GaAs
					—	**30.2**	4-terminal mechanically stacked
	8/90	0.0634	973	1416	83.8	**22.9**	InP top cell, 50 suns
		0.0663	445	1321	75.7	**8.9**	0.75-eV GaInAs bottom cell
			—	—	—	**31.8**	3-terminal InP/GaInAs tandem
Spire	5/91	0.250	1154	4988	86.4	**27.6**	GaAs, 180 suns
	5/91	0.250	1065	5911	80.2	**21.3**	GaAs on Si, 237 suns
Boeing	10/89	0.053				**32.6**	GaAs/GaSb, 100 suns
							4-terminal mechanically stacked
Varian/Stanford/Sandia	9/88	0.317				**29.6**	GaAs/Si, 350 suns
							4-terminal mechanically stacked
Submodules							
Boeing	3/93	41.4				**25.1**	3 mech. stacked GaAs/GaSb units

Notes: The independently confirmed efficiencies in this table are [8] or from the authors laboratory at NREL. All measurements were performed under standard reporting conditions of 25°C, 1000 W/m^2, ASTM E892 global spectrum. The area definition used for non-concentrator cells is total area. For concentrator cells the area is the designated illumination area (total are minus peripheral bus bars). The reference spectrum for concentrators is the ASTM E891 direct normal reference spectrum and its one-sun is arbitrarily taken to be 1000 W/m^2.

Table 6.4: Summary of confirmed cell efficiencies for amorphous silicon-based technologies.

	Date	Area (cm²)	Voc (mV)	Jsc (mA/cm²)	FF (%)	Efficiency (%)	Structure/Comments
cells							
ARCO	1/89	3.960	874	15.62	71.3	**9.7**	a-Si, not stabilized
APS	1/91	0.998	872	16.54	71.2	**10.3**	a-Si, not stabilized
Chronar	10/90	1.060	864	16.66	71.7	**10.3**	a-Si, not stabilized
ECD	2/88	0.268	940	15.20	69.4	**9.9**	ITO/a-Si/ss, not stabilized
	2/88	0.268	2541	6.96	70.0	**12.4**	ITO/aSi/aSiGe/ss, unstabilized
USSC/Cannon	1/92	0.284	1621	11.72	65.8	**12.5**	ITO/a-Si/a-SiGe/ss, not stabilized
IEC	11/87	0.284	862	17.60	65.8	**10.0**	Photo CVD a-Si, not stabilized
Glasstech	9/89	0.986	886	17.46	70.4	**10.9**	a-Si, not stabilized
Sanyo	4/92	1.00	887	19.4	74.1	**12.7**	a-Si, not stabilized
Solarex	4/87	1.077	879	18.80	70.1	**11.5**	a-Si, not stabilized
Spire	10/87	0.758	1685	9.03	68.1	**10.3**	a-Si/a-Si:Ge, not stabilized
	12/86	0.099	878	16.60	72.2	**10.5**	a-Si, not stabilized
Sharp	12/92	1.00	2289	7.9	68.5	**12.4**	a-Si/a-Si/a-SiGe, not stabilized
modules							
Sanyo	12/92	100	12.5V	1.3	73.5	**12.0**	a-Si, not stabilized
USSC	12/93	903	2.32V	6.47A	61.2	**10.2**	a-Si/a-Si/a-SiGe, "stabilized"
	12/93	906	2.40V	6.57A	67.5	**11.8**	a-Si/a-Si/a-SiGe, not stabilized

Notes: The independently confirmed efficiencies in this table are from [8] or NREL. All measurements were performed under standard reporting conditions of 25°C, 1000 W/m², ASTM E892 global. The area definition used for non-concentrator cells is total area. The aperture area (total area minus frame area) is used for modules.

Significant advancement in the a-Si technology is represented by the recent progress in stabilized efficiencies of modules. The stabilized efficiency is determined after continuous light soaking under AM1.5 conditions for 2000 h, with the device not exhibiting any further decrease in output. A triple-junction (a-Si/a-Si/a-Si:Ge), 962 cm^2 aperture area prototype by Solarex stabilized at 7.2%. A two-junction (a-Si/a-Si) USSC production module (3676 cm^2 aperture area) stabilized at 6.3%, and recently, a triple-junction (a-Si/a-Si:Ge/a-Si:Ge) was verified with a stabilized efficiency of 10.2%. These compare to 3%–4% efficiencies encountered 4–6 years ago. Current module durability is indicated by the ability of commercial modules produced by APS, UPG, and USSC to pass the NREL interim qualification tests and procedures for thin-film modules [27]. Certainly, the industrial component of this technology has shown its commitment with the initiation of two 10 MW/yr production lines for multijunction modules. However, the aggressive goal to demonstrate 10–12% stabilized modules within 2 years will be a major indicator for the future of this technology. Without some indication of stable, long-term operation with reasonable output, confidence in this photovoltaic approach will not be realized.

6.3.5.2 *Polycrystalline technologies*

Improvements in cost and performance may also be attained by utilizing other semiconductors for solar cells. The arguments for thin-film approaches for terrestrial photovoltaics are primarily based on materials utilization (less material needed and more efficient module area coverage), large-scale production potential, and lower energy consumption for producing product. Each period of time seems to highlight one of the photovoltaic cell technologies because of some extraordinary advancement in performance. This period belongs to two thin-film options, CuInSe$_2$ and CdTe.

6.3.5.2.1 Copper indium diselenide and related materials

Interest in Cu-ternary semiconductors for possible solar cell application began in the early 1970s, precipitated by the limitations that existed with both Si and Cu$_2$S cells. Of the options, CdS/CuInSe$_2$

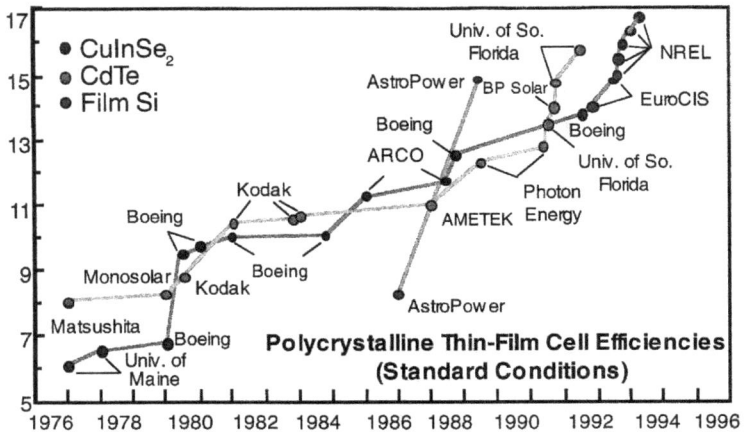

Figure 6.9:   Evolution of CIS, CdTe, and film Si technologies.

was demonstrated as a viable cell option, first in single-crystal then in thin-film form. A time scale portraying the evolution of this technology is presented in Fig. 6.9.

Two positive factors can be cited in favor of this thin-film technology. The first is its inherent stability. Unencapsulated cells have been tested for extended periods of time, with and without illumination, and show no degradation. Encapsulated modules have been tested and monitored outdoors for periods exceeding 3 years with no change in operating characteristics. A second positive indication for this technology is that it is possible to produce larger-area modules ($\sim$900 cm^2) with essentially the same conversion efficiencies as the better research cells — in excess of 11%, and recently almost 13% [28]. This gives some confidence that the process is scalable — a definite requirement for any of the photovoltaic approaches. Table 6.5 provides a summary of these devices.

This demonstration that the performances are not limited to laboratory or research cells, but can be produced in commercial-scale modules, is one of the more significant developments in the evolution of thin-film photovoltaics.

The structure of this device is basically that of a heterojunction. Various compositions and device structures are used to optimize

Table 6.5: Summary of confirmed efficiencies for CuInGaSe based polycrystalline, thin-film technologies.

	Date	Area (cm²)	V_{OC} (mV)	J_{SC} (mA/cm²)	FF (%)	Efficiency (%)	Structure/ Comments
cells							
ARCO	6/88	2.400	464	38.5	65.4	**11.7**	ZnO/[Cd,Zn]S/CuInSe₂
Siemens	11/92	0.192	539	33.7	73.6	**13.4**	ZnO/CdS/CuInSe+(I-II-VI₂) multinary/Mo
Boeing	12/86	1.074	509	30.4	66.0	**10.2**	CdS/CuInSe₂
IEC	6/93	1.003	604	34.3	67.9	**14.1**	ZnO/[Cd,Zn]S/CuInGaSe₂
ISET	9/87	1.028	445	35.0	64.6	**10.1**	ZnO/[Cd,Zn]S/CuInSe₂
	191	0.994	483	35.6	66.7	**11.5**	ZnO/CdS/CuInSe₂/Mo/Glass
NREL	3/96	0.414	674	34.0	77.3	**17.7**	MgF₂/ZnO/CdS/CuInGaSe₂
Solarex	11/94	0.414	594	32.8	73.8	**14.4**	CdS/CuInGaSe₂/Mo/Glass
modules							
ARCO	5/91	3883	24.0V	0.244A	64	**9.7**	ZnO/CdZnS/CuInSe₂
	1/89	3985	23.0V	2.46A	60	**8.5**	ZnO/CdZnS/CuInSe₂
	6/88	938	25.4V	0.637A	64	**11.1**	ZnO/CdZnS/CuInSe₂
EPV	10/94	202	11.5V	0.248A	64	**9.0**	ZnO/CdS/CuInSe₂/Mo/Glass
Siemens Solar	4/94	69.1	7.46V	0.173A	68	**12.7**	ZnO/CdZnS/CuIn,GaSe₂
	7/94	3859	27.2V	2.40A	61	**10.3**	ZnO/CdZnS/CuIn,GaSe₂
ARCO	6/88	938	25.9V	0.637A	64	**11.1**	ZnO/CdZnS/CuIn,GaSe₂
ISET	5/93	846	29.7V	0.284A	57	**5.7**	ZnO/CdZnS/CuInSe₂
Boeing	12/86	97.0	1.78V	0.774A	64	**9.1**	ZnO/CdZnS/CuInSe₂

Notes: The independently confirmed efficiencies in this table are from [8] or from the author's laboratory at NREL. All measurements were performed under standard reporting conditions of 25°C, 1000 W/m², ASTM E892 global. The area definition used for non-concentrator cells is total area. The aperture area (total area minus frame area) is used for modules.

cell performance, as indicated in Fig. 6.9. Two recent reports have further increased interest in this technology. The first is the EuroCIS (joint Sweden, France, and Germany) report of a 16.9% *active-area* (0.33 cm^2 area) cell produced by vacuum deposition [29]. After concurrent light soaking at an elevated temperature, this same cell recently reached a reported 17.5% active-area efficiency. The second is a ZnO/(Cd,Zn)S/Cu(In,Ga)Se$_2$ [CIGS] (vacuum-deposited) cell by Boeing, having a total-area efficiency of 14.7% [30]. The *highest confirmed efficiency* (under standard conditions) for these type of cells is the 17.1% MgF$_2$/ZnO/CdS/Cu(In,Ga)Se$_2$/Mo device (0.413 cm^2 area) produced by NREL [31].

6.3.5.2.2   Cadmium telluride

This is certainly not a new semiconductor in the photovoltaic literature; CdTe has been receiving attention as a photovoltaic candidate since the early 1960s. Although its 1.5 eV bandgap is well matched to the terrestrial spectrum, the problems with this compound semiconductor have been materials related. It is difficult to make good, reliable, and reproducible electrical contacts, especially to the p-type material. The electrical properties are difficult to control. Also, it tends to be semi-insulating. Metallic dopants such as Cu (that do lower the resistivity) have very high diffusion coefficients. The cells and modules have shown some instabilities in operation [32]. A number of efficiency highlights (Table 6.6) have kept interest in this technology. The 15.8%, 1 cm^2 cell, fabricated by the University of South Florida (using close-space vapor transport for the CdTe and chemical bath deposition for the CdS) stands as the highest efficiency for a thin-film polycrystalline cell technology at this writing [33]. Results have recently been reproduced at South Florida by a new group and at NREL, giving further support to this approach. Solar Cells, Inc., is currently using the closed-space process for both heteropartners to produce 0.09 m^2 area modules. In addition, the development of 8.1% (900 cm^2) and ∼5% (3600 cm^2) CdTe modules by Golden Photon, in which a low-cost spraying process is used for the semiconductor components, has increased confidence in the potential of this photovoltaic approach for two reasons. First, it is

Table 6.6: Summary of confirmed efficiencies for CdTe containing polycrystalline, thin-film technologies.

	Date	Area (cm²)	Voc (mV)	Jsc (mA/cm²)	FF (%)	Efficiency (%)	Structure/ Comments
cells							
IEC	10/91	0.191	790	20.10	69.4	**11.0**	Glass/ITO/CdS/CdTe/Cu/Au
NREL	7/95	0.824	840	20.66	74.0	**12.8**	Glass/SnO₂/CdS/CdTe
Photon Energy	5/89	0.313	783	24.98	62.7	**12.3**	Glass/SnO₂/CdS/CdTe
	5/91	0.300	788	26.18	61.4	**12.7**	Glass/SnO₂/CdS/CdTe
SMU	4/88	1.022	736	21.90	65.7	**10.6**	Glass/SnO₂/CdS/CdTe/HgTeGa
Univ.S.Florida	6/92	1.047	843	25.09	74.5	**15.8**	MgF₂/7059 Glass/SnO₂/CdS/CdTe/C/Ag
AMETEK	10/89	1.068	767	20.93	69.6	**11.2**	Glass/SnO₂/CdS/CdTe/ZnTe/Ni
Georgia Tech.	6/91	0.080	745	22.10	66.0	**10.9**	Glass/SnO₂/CdS/MOCVD CdTe
Solar Cells Inc.	4/95	1.115	828	20.9	74.6	**12.9**	MgF₂/Glass/SnO₂/CdS/CdTe
U.Toledo	11/92	0.114	815	17.61	72.8	**10.4**	Glass/SnO₂/CdS/Sputtered CdTe/Cu/Au
modules							
AMETEK	10/88	100	9.63V	0.125A	57	**6.8**	
Photon Energy	9/91	832	~21V	0.573A	55	**8.1**[†]	
Golden Photon	8/93	3528	44.2V	1.10A	57	**7.7**[††]	
Solar Cells Inc.	5/93	63.6	6.62V	0.137A	69	**9.8**	
	2/95	6693	92.8V	0.966A	66	**8.6**[†††]	

[†]Measured outdoors in Golden, CO, 1013-W/m² total irradiance, 33°C module temperature
[††]Measured outdoors in Golden, CO, 1013-W/m² total irradiance, 31°C module temperature
[†††]Measured outdoors in Golden, CO, 1032-W/m² total irradiance, 24°C module temperature

Notes: The independently confirmed efficiencies in this table are from [8] or from the author's laboratory at NREL. All measurements were performed under standard reporting conditions of 25°C, 1000 W/m², ASTM E892 global. The area definition used for non-concentrator cells is total area. The aperture area (total area minus frame area) is used for modules.

shown that it is possible to produce efficient, large-area modules using a very inexpensive production method. Second, the outdoor stability tests of these prototype CdTe modules indicates that >15-year lifetimes are achievable.

These results on $CuInSe_2$ and CdTe are impressive. It should be emphasized that the high efficiencies have been reached on laboratory-scale, relatively small-area, research cells and that the long-term stability of these material systems to the full range of qualification and accelerated testing is an active area or research. Polycrystalline thin-film cells require further development to achieve the >15% efficiencies projected to meet the 1995 DOE thin-film module goals. For both these technologies, concerns for the environmental impact of the Cd, Se, and Te are receiving scrutiny. Investigations of the potential effects of processing and using these materials on health and safety are underway, but preliminary indications are that no adverse environmental problems should exist under usual processing and operating conditions [34–36]. However, these environmental concerns will continue to receive attention by those wanting to ensure a safe, clean energy source.

6.3.5.2.3   Advanced approaches

Photovoltaics remains an evolving technological field, and new or redefined approaches are continually being investigated and tested to find better and less costly devices, components, and systems. Several cell types have recently been demonstrated that are interesting from cost and performance aspects. These include the use of new structures (quantum-well solar cells) [4], the use of new materials (iron sulfide and other pyrites) (Ennaoui and Tributsch, 1984), and non-conventional approaches (nanocrystalline thin films).

In this last category is the recently announced electrolytic cell, termed the Grätzel cell [37]. It uses a spectrally sensitized, thin ceramic membrane. This film consists of nanometer-sized, colloidal titanium dioxide particles sintered to permit lateral charge transport. Such membranes have been reported to yield remarkable efficiencies, exceeding 90% for certain transition metal complexes within the energy range of their absorption bands. The cell is encapsulated between two glass plates, coated with a thin conducting SnO_x layer. The titanium dioxide film, which has nanocrytalline grain structure and acts as the absorber, is deposited onto one glass substrate. The light is absorbed by a monomolecular layer of a special transition metal complex (the sensitizer, usually using a ruthenium bipyridyl complexes). Under illumination, the sensitizer injects an electron into the titanium dioxide conduction band. The electrons pass through the collector contact into the external circuit to provide the current. These electrons return to the cell via the counter electrode. The sensitizer film is separated from the counter electrode by the electrolyte, which uses a redox system (e.g., iodine/iodine), which transports electrons from the counter electrode to the positively charged sensitizer layer. Modules with areas to $100\,cm^2$ have been tested, and unconfirmed efficiencies (reported to be measured under AM1.5) to the 10% level have been reported. It is interesting that the efficiency of the cell is reported to be higher under cloudy conditions, exceeding that of commercial silicon cells measured concurrently. NREL has recently evaluated a cell provided by Grätzel, with a 4.8% efficiency ($V_{oc} = 0.667\,V$, $J_{sc} = 10.73\,mA/cm^2$, FF = 67%) measured under standard conditions for this $0.604\,cm^2$ area device. [This cell was evaluated as part of a characterization intercomparison, and the cell was not one of the better devices.] Many questions and future areas of development remain for this device. These include the improvement of the spectral matching of the sensitizer absorption, the improvement of the conductive transparent contacts, and the confirmation of the performance of the cells and modules. In addition, major concerns with the stability of the cell, especially the dye and electrolyte, have been expressed from some members of the photovoltaic community. However, the cell certainly demonstrates a potentially low-cost path to the realization of a solar electric technology if the performance objectives are reached.

6.3.6 *Thermophotovoltaics*

A thermophotovoltaic (TPV) system is composed of a heat source, an emitting surface, spectral filter and photovoltaic cells (see [4] for details). Practical TPV heat sources include; combustion of flammable materials, solar concentrators, and nuclear sources. These sources operate in the 900–1800°C. In the terrestrial environment nitrogen oxides start to form at temperatures above 1400°C setting an upper temperature limit to combustion sources. At 1400°C the power density of a black body emitter is about $440,000 \, \text{W/m}^2$ (440 times the power density at 1-sun). Emitting surfaces include spectrally selective emitters that radiate in a narrow wavelength region more suitable for conventional photovoltaics, and gray body emitters that radiate with a spectrum similar to a black-body. These emitters include SiC and various rare earth oxides. It is desirable to reflect all infrared light with energies less than the bandgap energy of the PV cell back to the heat source. This can be accomplished with proper cell design (back surface reflector and minimum free carrier absorption), or through the use of a dichroic reflector placed in front of the TPV cells. TPV cells typically have a bandgap energy of less than 0.75 eV. The GaSb material system is popular because GaSb wafers are inexpensive and the cell can be fabricated using conventional diffusion technology. The GaInAsP or GaInAsSb material systems allow a wide range of bandgap energies to be explored. These devices have excellent photovoltaic properties and wide process control. These devices are grown on InP or GaSb substrates using liquid or vapor phase epitaxy. Small prototype TPV systems have been designed or developed for radionucleotide based deep space probes [38], air cooled 100W gas powered portable power source [39], 0.1-W candle powered source [40], multi-kilowatt solar TPV system [41], and a grid-independent residential gas furnace [42].

6.3.7 Acknowledgments

This work was performed, in part, under U.S. Department of Energy contract DE-AC02-83H10093 with the National Renewable Energy Laboratory.

6.3.8 References

[1] *Proceedings of the IEEE Photovoltaic Specialists Conference*, No. 18-24, IEEE New York, 1985–1994.
[2] *Solar Cells*, Vol. 1–24, Elsevier Sequoia, 1979–1994.

[3] *Progress in Photovoltaics*, Vol. 1-4, John Wiley and Sons, 1993–1996.

[4] *The First NREL Conference on Thermophotovoltaic Generation of Electricity*, 1994, American Institute of Physics Proc. Vol. 321. *The Second NREL Conference on Thermophotovoltaic Generation of Electricity*, 1995, AIP Proc. Vol. 358.

[5] U.S. Department of Energy (1991) Photovoltaics Program Plan FY 1991-FY 1995, DOE/CH10093-92, DE91002139. Updated plan under consideration, FY1996-FY2000.

[6] U.S. Department of Energy (1992) Solar 2000, Vol. 1 and 2, National Renewable Energy Laboratory, Golden, Colorado.

[7] K.A. Emery and C.R. Osterwald, *Current Topics in Photovoltaics* 3, 1988.

[8] M.A. Green, K. Emery, K.Bücher, and D. King, "Solar cell efficiency tables (Version 6)", *Prog. Photovolt.* 3, 1995, 229–233.

[9] PV News, February 1996.

[10] P.A. Iles and F. Ho, *Proceedings of the First World Conference on Photovoltaic Energy Conversion*, Hawaii, IEEE, New York, 1995, pp. 1957–1962.

[11] J. Zhao, A. Wang, and M.A. Green, *Prog. in Photovolt.* 2, 1994, 227–230.

[12] P.J. Verlinden, R.A. Crane, R.M. Swanson, T. Iwata, K. Handa, and D.L. King, *Proceedings of the 12th European Communities Photovoltaic Solar Energy Conference*, Kluwer Scientific Publ., The Netherlands, 1994.

[13] J. Zhao, A. Wang, M. Taouk, F. Yun, A. Ebong, A.W. Stephens, S.R. Wenham, and M.A. Green,*Proceedings of the 23rd IEEE Photovoltaic Conference*, IEEE, New York, 1993, pp. 1246–1249.

[14] T.M. Bruton, K.C. Heasman, J.P. Nagle, and R.R. Russell, *Proceedings of the 23rd IEEE Photovoltaic Conference*, IEEE, New York, 1993, pp. 1250–1251.

[15] N. Narayanan, J. Zolper, F. Yun, S.R. Wenham, A.B. Sproul, C.M. Chong, and M.A. Green, *Proceedings of the 21st IEEE Photovoltaic Specialists Conference.*, IEEE, New York, 1990, pp. 678–680.

[16] T.L. Jester, *Proceedings of the 13th NREL Photovoltaics Program Review*, American Inst. Physics, Vol. 353, 1995, pp. 326–333.

[17] J.H. Wohlgemuth, *Proceedings of the 13th NREL Photovoltaics Program Review, American Inst. Physics*, Vol. 353, 1995, pp. 326–333.

[18] A.M. Barnett, W. Bloothood, W.R. Bottenberg, S.R. Collins, D.H. Ford, R.B. Hall, C.L. Kendall, S.M. Lampo, J.A. Rand, and A.M. Trabant, *Proceedings of the 12th European Communities Photoboltaic Solar Energy Conference*, Kluwer Scientific, 1994.

[19] J.A. Rand, A.M. Barnett, J.C. Checchi, S.R. Collins, D.H. Ford, R.B. Hall, C.L. Kendall, S.M. Lampo, and A.M. Trabant, *Proceedings of the 13th NREL Photovoltaics Program Review, American Inst. Physics*, Vol. 353, 1995, pp. 283–289.

[20] J.D. Levine, G.B. Hotchkiss, and M.D. Hammerbacher, *Proceedings of the 22nd IEEE Photovoltaic Specialists Conference*, IEEE, New York, 1991, pp. 1603–1606.

[21] R. Schmidt, *Proceedings of the 23rd IEEE Photovoltaic Specialists Conference*, IEEE, New York, 1994, pp. 1078–1081.

[22] J.M. Olson, S.R. Kurtz, A.E. Kibbler, and P. Faine, *Proceedings of the 21st IEEE Photovoltaic Specialists Conference*, IEEE, New York, 1990, pp. 24–29.

[23] J.C.C. Fan, R.W., McClell, and B.D. King, *Proceedings of the 17th IEEE Photovoltaic Specialists Conference*, IEEE, New York, 1984, pp. 31–35.

[24] J.E. Avery, L.M. Fraas, V. Sundaram, D.J. Brinker, and M.J. O'Neill, *Proceedings of the* 21st IEEE Photovoltaic Specialists *Conference*, IEEE, New York, 1990, pp. 1277–1279.

[25] M.W. Wanlass, T.J. Coutts, J.S. Ward, K.A. Emery, T.A. Gessert, and C.R. Osterwald, *Proceedings of the 22nd IEEE Photovoltaic Specialists Conference*, IEEE, New York, 1992, pp. 38–45.

[26] K.A. Bertness, S.R. Kurtz, D.J. Friedman, A.E. Kibbler, C. Kramer, and J. Olson, *Proceedings of the First World Conference on Photovoltaic Energy Conversion*, Hawaii, IEEE, New York, 1995, pp. 1671–1678.

[27] R. DeBlasio, L. Mrig, and D. Waddington, Interim Qualification Tests and Procedures for Terrestrial Photovoltaic Thin-Film Flat-Plate Modules, Technical Report SERI/TR-213-3624, 1990.

[28] R. Gay, *Proceedings of the 12th European Communities Photovoltaic Solar Energy Conference*, Kluwer Scientific Publ., The Netherlands, 1994, p. 935.

[29] D. Schmid, M. Ruckh, G. Grunwald, and H.W. Schock, *J. Appl Phys.* 73 (1992) 2902. Also, M. Bodegård, L. Stolt, and J. Hedström, *Proceedings of the 12th European Communities Photovoltaic Solar Energy Conference*, Kluwer Scientific Publ., The Netherlands, 1994, pp. 1743–1753.

[30] W.S. Chen, J.M. Stewart, W.E. Devaney, R.A. Mickelson, and B.J. Stanbery, *Proceedings of the 23rd IEEE Photovoltaic Specialist Conference*, IEEE, New York, 1993, pp. 422–425.

[31] J.R. Tuttle, M.A. Contreras, T.J. Gillespie, K.R. Ramanathan, A.L. Tennant, J. Keane, A.M. Gabor, and R. Noufi, *Prog. Photovolt.* 3, 1995, 235–238.

[32] L. Mrig (ed.), *Photovoltaic Performance and Reliability Workshop*, NREL/CP-410-6033 DE94000236, NREL, Golden, Colorado, September 1993, 8–10.

[33] T.L. Chu, S.S. Chu, C. Ferekides, J. Britt, and C. Wang, *J. Appl. Phys.* 70, 1991, 7608.

[34] M.H. Patterson, A.K. Turner, M. Sadeghi, and R.J. Marshall, *Proceedings of the 12th European Communities Photovoltaic Solar Energy Conference*, Kluwer Scientific Publ., The Netherlands, 1994, pp. 948–951.

[35] K.M. Hynes, A.E. Baumann, and R. Hill, *Proceedings of the 12th European Communities Photovoltaic Solar Energy Conference*, Kluwer Scientific Publ., The Netherlands, 1994, p. 309.

[36] P. Moskowitz, K. Zweibel, and M.P. DePhillips (eds.), *Proceedings of the Understanding and Managing Health and Environmental Risks of CIS, CGS, and CdTe PV Module Production and Use*, Brookhaven National Labs Tech. Rep. 61480, NREL, Golden, Colorado, April, 1994.

[37] B. O'Regan, J. Moser, M. Anderson, and M. Grätzel, *J. Phys. Chem.* 94, 1990, 8720; Also, A. Kay and M. Grätzel, *J. Phys. Chem.* 97, 1993, 6272.

[38] A. Schock, C. Or, and V. Kumar, *Proceedings of the Second NREL Conference on Thermophotovoltaic Generation of Electricity*, AIP Proc. Vol. 358, 1995, pp. 81–97.

[39] L.M Fraas, and D.J. Williams, *Proceedings of the the Second NREL Conference on Thermophotovoltaic Generation of Electricity* AIP Proc. Vol. 358, 1995, pp. 128–133.

[40] L.M. Fraas and D.J. Williams, *Proceedings of the the Second NREL Conference on Thermophotovoltaic Generation of Electricity*, AIP Proc. Vol. 358, 1995, pp. 134–137.

[41] K.W. Stone, D.L. Chubb, D.M. Wilt, and M.W. Wanlass, *Proceedings of the Second NREL Conference on Thermophotovoltaic Generation of Electricity*, AIP Proc. 358, 1995, pp. 199–209.

[42] R.E. Nelson, *Proceedings of the Second NREL Conference on Thermophotovoltaic Generation of Electricity*, AIP Proc. Vol. 358, 1995, pp. 221–237.

6.4   Discussion after Keith Emery's presentation

Faiman: You mentioned a theoretical calculation for a double junction a-Si cell in which the optimal band gaps were 2 eV for the upper layer and 1.5 eV for the lower layer. Is this a general result or does it depend on the electronic structure of silicon?

Emery: The calculation was published by Carlson et al of Solarex and I believe it appears in the Proceedings of one of the IEEE PV Specialist Conferences in recent years. The starting assumption was for a mid-cell material with a 1.7 eV bandgap, that is, a-Si. Carbon doping is used to increase the bandgap of the upper layer and germanium doping to decrease that of the base material. For a mid-cell material having a different band gap the optimal combination for high efficiency might turn out to be a different set of values.

Cahen: You devoted much time to the question of the stability of a-Si but did not mention the stability of either thin-film polycrystalline cells or thermophotovoltaic cells.

Emery: In the case of thin-film polycrystalline cells the first generation of modules have begun to undergo the kind of rigorous qualification tests required of conventional modules. As yet there are no startling revelations. For example, one might not be surprised if the diffusion of impurities were found to cause problems in these materials (referring to all thin-film types) which are, after all, only a few microns thick. But field tests have not indicated any such problem, admittedly, after less than 5 years of experience.

There is however a major obstacle, even in the case of single-crystal silicon, in stating what the expected module life-time should be. On the one hand there is no comprehensive list of possible degradation mechanisms — new ones are still being discovered. Secondly, there is not yet a large enough data base of observed causes of failure. It is, therefore, not even possible to claim a 30 expected year life-time for these modules. We need more data.

As for thermophotovoltaic cells, this technology is still at the prototype cell scale. No large area devices have yet been made let alone subjected to qualification tests.

Weiser: Aren't these really Peltier-effect devices?

Emery: No. They are genuine photovoltaic devices which detect infrared photons emitted by low temperature sources — typically at $1000°C$ rather than by a $5800\,K$ sun. The band gap is accordingly small.

Question: What kind of efficiency do they have?

Emery: That question cannot be answered in a meaningful manner since, unlike the solar cell case, there are no standard reference conditions. In a deep space probe, for example, the heat source may be a relatively low temperature radio-nuclide. On the other hand, for terrestrial purposes it may be a solar furnace.

What I can say is that, for the cells that have been fabricated to date, both the open-circuit voltage and the fill factor are reasonably close to their theoretical limits. Perhaps a more meaningful practical figure of merit would be the extent to which a thermophotovoltaic cell might be more or less efficient than a Stirling engine for a given heat source. In this respect, it is interesting to note that much of the research funding for thermophotovoltaics is coming from the Gas Research Institute. They are interested in methods of increasing the combustion efficiency of gas-fired power stations.

6.5 Quantum well solar cells (Professor Keith Barnham)

A keynote lecture presented by Keith Barnham (Blackett Laboratory, Imperial College of Science, Technology and Medicine, London, SW7 2BZ, UK). Please note that although this invited keynote lecture was presented by Professor Barnham, the paper that he provided for publication in the Symposium Proceedings — which is reproduced here — was co-authored by: Ian Ballard, Jenny Barnes, James Connolly, Paul Griffin, Enrique Grünbaum (University of Tel Aviv, currently visiting Department of Materials, University of Oxford, Parks Road, Oxford OX1 3PH), Benjamin Kluftinger, Jenny Nelson, Ernest Tsui, and Alexander Zachariou.

6.5.1 *Abstract*

Quantum Well Solar Cells (QWSCs) are an alternative to tandem or multi-bandgap approaches to the problem of the efficiency limits imposed on conventional solar cells by the single, fixed bandgap. Though quantum wells have been extensively studied for lasers and other opto-electronic applications, our work has been the first study of the use of quantum wells in solar cells. In this paper we discuss quantum wells and the advantages they have in enhancing solar cell efficiency. We present recent experimental results that show that the open-circuit voltage (V_{oc}) is enhanced over that of a comparable conventional cell formed from the well material, by more than the change in the absorption edge. We also discuss theoretical and experimental studies which seek to determine the quasi-Fermi level separation in the QWSC. The variation of the quasi-Fermi levels in the depletion region is fundamental to understanding the voltage enhancement. We also discuss the optimization of quantum well solar cells in the AlGaAs/GaAs and the GaAs/InGaAs material systems. In the latter case the main problem is the incorporation of a large number of wells in a strained system. Another material system, InP with InGaAs wells, has wells which are lattice matched and relatively deep. Our success in overcoming background doping problems in this system will be reported, resulting, for the first time, in a QWSC which has efficiency enhanced with respect to a comparable control cell made from the well material. We also present some preliminary new results in the InP/InGaAs system, which show that QWSCs with deep wells have a better variation of efficiency with temperature than conventional cells. This will prove important for light concentration applications.

6.5.2　*Introduction*

The fundamental limitation on the efficiency of a solar cell is a result of the solar radiation having a broad spectrum of energies and a conventional cell having one fixed bandgap energy below which photons of light are not absorbed. The choice of semi- conductor material is a trade-off between maximizing current output with a low bandgap and voltage output with a high bandgap semi-conductor. As a result the maximum theoretical efficiency [1] of a conventional solar cell is just above 30% in unconcentrated sunlight at a bandgap just below that of GaAs. Tandem and multi-bandgap cascade approaches have the potential to achieve higher efficiencies. Henry [1] has calculated that a 3 cell cascade would have an upper efficiency limit of 56% in 1000× concentrated AM1.5 sunlight.

The QWSC is a novel device [2] with the potential to achieve high efficiency in an alternative approach to tandem or cascade systems.

In its simplest form (shown in Fig. 6.10(a) in comparison with a conventional solar cell in Fig. 6.10(b)) it consists of a multi-quantum well (MQW) system in the undoped region of a p-i-n solar cell. For light with energy greater than the bandgap (E_g) the quantum-well (QW) cell is expected to behave like a conventional solar cell.

However, light with energy below E_g and above E_a can be absorbed in the quantum wells. Our studies show that if the material quality is good the photogenerated electrons and holes escape from the wells and increase the output current [3]. The output voltage lies between that of a conventional cell formed from the material of the well and one formed from the barrier material [4]. In AlGaAs/GaAs test devices we have obtained efficiency enhancements of a factor of

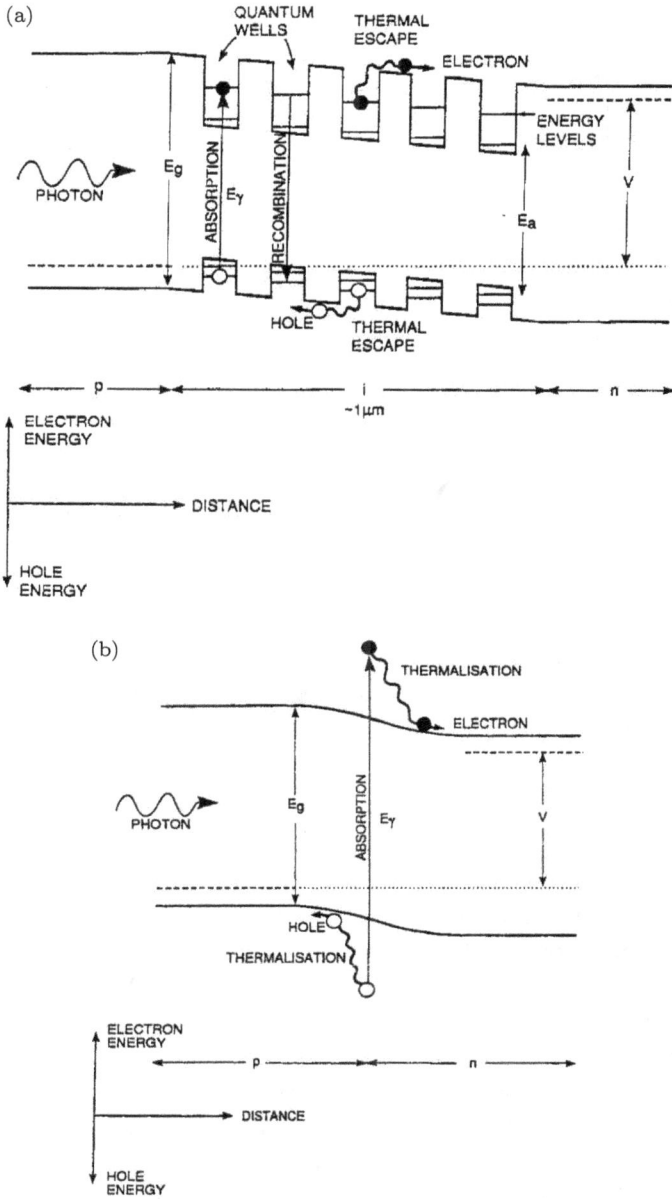

Figure 6.10:   (a) Schematic diagram of a quantum well solar cell. (b) Schematic diagram of a conventional, single bandgap solar cell.

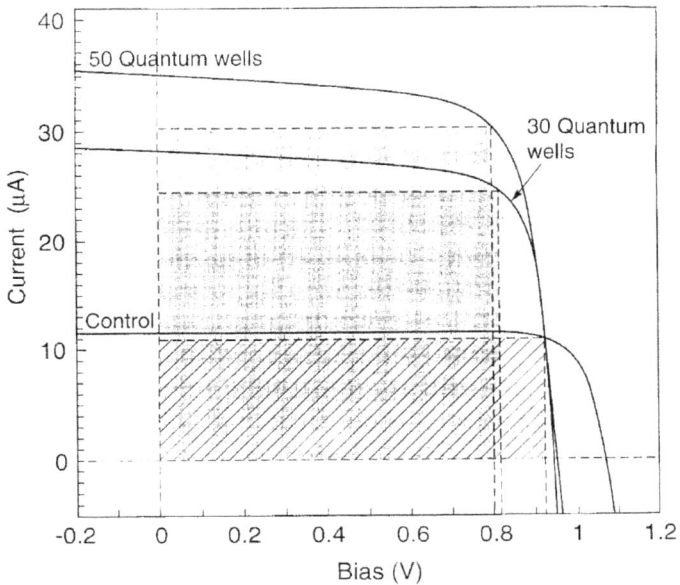

Figure 6.11: Current–voltage characteristics under white light illumination (3000 K spectrum) for test devices with 30 and 50 quantum wells and a control device without wells but with the same i-region thickness as the 30 well sample. The output powers are given by the areas of the rectangles.

more than two in a white light source when cells with quantum wells are compared with identical cells without wells [3]. Typical examples are shown in Fig. 6.11.

Quantum wells have been extensively studied for lasers and other optoelectronic devices. Our work represents the first study of their use in solar cells. In Section 6.5.3 of this review, we give a brief introduction to quantum wells and, in Section 6.5.4, we will discuss their relevance to the enhancement of solar cell efficiency.

Reviews of our early work on QWSCs can be found in [4, 5]. In this paper, we will concentrate on recent results. As there have been few studies of MQW systems in the situation where power can be extracted we have had to undertake many fundamental studies ourselves. Sadly we will not have space to discuss all of these. Experimental and theoretical studies of carrier escape by steady-state photoconductivity and photoluminescence can be found in [6, 7] and by time resolved photoluminescence in [8]. The theory of the QWSC has been recently reviewed in [9].

Others, have confirmed our observation that quantum wells in the AlGaAs/GaAs system enhance the short-circuit current (I_{sc}) sufficiently to overcome the voltage loss and ensure efficiency enhancement [10]. However, there has been considerable controversy surrounding the effect of the wells on the open-circuit voltage (V_{oc}). It has been argued on the basis of generalized detailed balance [11] that an ideal quantum well solar cell would have a voltage behavior identical to that of an ideal conventional cell with the bandgap of the well material. In Section 6.4, we discuss some new experimental results which show that in real material, in three different systems, the open-circuit voltage is above that of comparable homojunction cells constructed from the material in the well.

When a solar cell is illuminated the carriers in the depletion region are not in equilibrium and so cannot be described by a Fermi level. However the electrons can be considered to be in thermal equilibrium with each other and so can the holes. Hence it is usual to describe the electron and hole distributions in the depletion region by two quasi-Fermi levels. To fully understand the voltage dependence of the QWSC one needs to know how these quasi-Fermi levels vary in the quantum well region. In Section 6.5.6, we outline our recent experimental and theoretical work to establish the quasi-Fermi level separation.

Early studies of the QWSC were mainly with AlGaAs/GaAs material which is the best understood MQW system. In Section 6.5.7, we will discuss our progress in optimizing a QWSC in this system. The AlGaAs alloy, which forms the conventional parts of this device, is not an ideal solar cell material. Minority carrier lifetimes are poor and the bandgap is larger than the optimum. We have therefore extended our studies to other MQW systems which offer the prospect of higher efficiency. In one of these, a GaAs solar cell with InGaAs wells, the material in the well is strained to match the material in the barriers. This leads to particular problems which we are tackling as discussed in Section 6.5.8.

In another material system, InP with InGaAs wells, the wells are lattice matched and relatively deep. Our success in overcoming the background doping problem is discussed in Section 6.5.9. Since these wells are particularly deep they should lead to a large efficiency enhancement and have important implication for concentrator applications. Concentration systems can reduce the costs of solar electricity considerably. Mirrors and lenses are considerably cheaper than solar cells. One drawback, particularly at high concentration, is the fact that the cells get very hot and the efficiency falls as the bandgap shrinks with increasing temperature. However, as discussed in Section 6.5.10, QWSCs should have better temperature coefficient of efficiency than conventional cells and we will discuss preliminary results which demonstrate this effect. Section 6.5.11 contains our conclusions.

6.5.3   *Quantum wells*

Most physics and chemistry undergraduates will have encountered the potential well as their first problem in undergraduate quantum

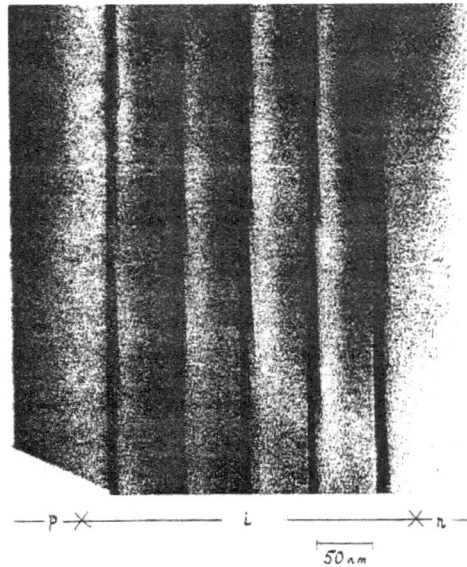

Figure 6.12:   Transmission electron micrograph of GaAs QWSC with 5 strained $In_{0.16}Ga_{0.84}As$ quantum wells grown in a p-i-n structure (QT422A).

mechanics. In the last decade or so, the new layer-by-layer crystal growth techniques, such as Metal-organic Vapor Phase Epitaxy (MOVPE) and Molecular Beam Epitaxy (MBE), have made it possible to produce such potential wells in practice.

A quantum well is formed when a very thin layer (around 1–20 nm, i.e., from a few to tens of atomic separations) of a narrower bandgap semiconductor (the well) is grown in-between regions of a wider bandgap semiconductor (the barrier). In good quantum wells the transitions between the two materials can be extremely sharp, changing over one or two atomic spacings. In Fig. 6.12, we show a transmission electron micrograph of one of our 5 well GaAs/InGaAs strained QWSCs.

Electrons and holes show all their quantum particle behavior in these wells. They can only occupy certain energy levels in the wells They have wave functions as shown in a schematic diagram of the conduction and valance band structure in Fig. 6.13. They can tunnel or be excited thermally out of the wells. They can be bound in pairs by Coulomb attraction to form relatively long-lived, hydrogen-like states known as *excitons*.

The structure in Fig. 6.10(a) is essentially similar to a MQW photodiode or modulator device, which operates in reverse bias, and to a quantum well laser which operates in the high forward-bias/forward current quadrant of the *I–V* characteristic beyond flat band. These systems have been very extensively studied in the past decade or so. Many of the latest CD players use quantum well lasers and the ultra-fast transistors used in satellite TV dishes have electrons moving along similar wells. In future the nonlinear optical properties associated with the excitons in quantum wells will enable MQWs to form the basis of the logic elements in all-optical computing systems.

Our work represents the first study of MQWs in the small-forward bias, reverse current region where power can be extracted from the device. It should be noted that MOVPE is one of the main methods used commercially to grow conventional cells for space applications. Once the development costs have been paid quantum wells could be

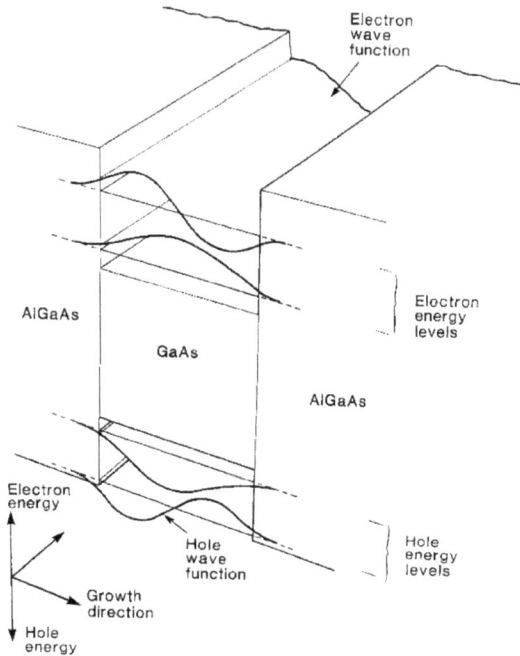

Figure 6.13:   Schematic diagram representing the conduction and valance band-edges in an AlGaAs/GaAs quantum well system and the allowed electron and hole wave functions.

incorporated for minimal extra cost. The price of GaAs substrates is falling fast and the costs of these types of cells can be further reduced when used in light concentration systems, as discussed in Section 6.5.10.

6.5.4   *Quantum wells and solar cells*

A conventional solar cell consists of a p-n junction in a semiconductor of bandgap E_g (Fig. 6.10(b)). Electron-hole pairs are produced when photons of light with energy E_γ greater than E_g are absorbed. If the photon is absorbed in the p–n junction region, or within a minority carrier diffusion length of it, the carriers are separated by the electric field and move in opposite directions, producing a current. Semiconductors with a smaller bandgap absorb more sunlight resulting in higher output *current*. However, the carriers rapidly lose energy and "thermalize" to the band-edges, so much of the incident photon energy ends up as heat. Using larger bandgap semiconductors reduces this energy loss and leads to higher output *voltage*. Since the output *power* is the product of current and voltage, the choice of bandgap is a compromise. This trade-off between current and voltage produces a maximum theoretical power conversion efficiency [1] for a conventional cell of about 30% at a bandgap close to that of GaAs (1.4 eV).

Higher efficiencies can be achieved by stacking together several solar cells with different bandgaps with the highest bandgap cells receiving the sunlight first. Connecting them in series, optically and electrically, is not simple. However, a high efficiency of 29.5% has recently been achieved for a GaInP/GaAs tandem [12].

The QWSC (Fig. 6.10(a)) is an alternative approach which works on a different principle. The region over which the electric field of the p–n junction operates is larger. It contains a layer of undoped (intrinsic) material and includes a number of quantum wells (ideally 50 or more). For photons of light with energy E_γ greater than E_g the new cell behaves like a conventional solar cell. However, many photons with energy less than E_g and greater than E_a will be absorbed in the quantum wells. The electrons and holes which are created in the well can either escape from the well (by tunneling or absorbing heat energy) or they will be lost by recombination. In perfect material this would be by radiative recombination. However, in practice, in presently available material, there is a considerable contribution from non-radiative recombination through trap states. Our results show that in good material, in particular when the background doping in the intrinsic region is low so that the built-in field is maintained across the i-region [3], escape processes dominate over recombination processes at room temperature. The escaping carriers are separated by the built-in field and the cell therefore generates extra current from the photons absorbed with energy less than E_g. The output voltage is somewhere between that of a conventional cell formed entirely from the well material and one of bandgap E_g formed from the barrier material as will be discussed in the next section. The output power of the QWSC will be larger than that of the conventional cell formed entirely from the barrier material if the gain in short-circuit current exceeds the loss in open-circuit voltage.

The light produces a separation in the Fermi levels in the p and n regions and hence gives rise to the output voltage V. As discussed in Section 6.5.2, within the depletion region the electrons and holes are no longer in equilibrium but the electrons can be described by one quasi-Fermi level and the holes by another. Determining the way these quasi-Fermi levels vary in a MQW system and join together in the n and p regions is one of the major questions about the QWSC which will be discussed in Section 6.5.6.

Note that in Fig. 6.10(a) the opposite process is occurring to that in the conventional cell in Fig. 6.10(b). In the QWSC the carriers are being thermally activated to the bulk band-edge. This property will be particularly important for concentrator cell applications as will be discussed in Section 6.5.10.

It should be noted that there have been few studies of MQW systems in the low forward-bias/reverse-current quadrant where power can be extracted. The fundamental physics turns out to be interesting and relevant to many optoelectronic devices. However, material quality is of particular importance in this regime, as we are utilizing the built-in field of the wide i-region rather than a p-n junction.

6.5.5   *Voltage enhancement in QWSCs*

It has now been clearly established, by our own experiments [3, 13] and those of others [10, 14, 15] in a number of material systems, that the short-circuit current I_{sc} of the QWSC can be enhanced over a conventional, homojunction cell formed from the barrier material (i.e. with bandgap. E_g in Fig. 6.10(a)). However, there has been some controversy over the question of the output voltage of an ideal QWSC. The problem is in determining where the quasi-Fermi levels lie in a QWSC. Considerable effort in many laboratories over a number of years was necessary to clarify similar questions in quantum well lasers.

In the original proposal [2] it was foreseen that the voltage dependence of the QWSC would be somewhere between that of the well and barrier, that is, determined by a bandgap which would be some fraction of the barrier bandgap E_g. For illustrative purposes an. upper limit to the efficiency of the QWSCs was estimated by assuming this fraction was unity.

Subsequently, Corkish and Green [16], used a model in which there was no carrier transfer between the QWs and the conventional parts of the cell, but the wells and bulk cell were combined in parallel. They concluded that, in the limit of radiative recombination dominance, enhancement of QWSC efficiency above the limit for an ideal single bandgap cell was possible for narrow wells. This model and other theoretical approaches are critically reviewed in [9].

Araujo and Marti [11] applied generalized detailed balance arguments to the case where the quasi-Fermi levels are constant throughout the cell. In the case of a QWSC with barrier bandgap E_g greater than the bandgap of the maximum efficiency conventional solar cell they predict that, in the limit of radiative recombination dominance, the quantum wells enhance the efficiency but not the open-circuit voltage compared to a conventional cell of bandgap E_g However, for a QWSC with barrier bandgap equal to or below that of the ideal conventional cell they conclude that, in the limit of negligible non-radiative recombination, the maximum efficiency of the QWSC cannot exceed that of the ideal, single-bandgap cell.

Recently, Anderson [17] has produced a model of QWSC behavior in which he demonstrates that introducing ideal MQW material into the depletion region of a realistic solar cell can enhance efficiency if the wells are shallow.

The behavior of the quasi-Fermi levels in real QWSCs must be understood in order to clarify the voltage dependence. In the next section, we will outline how we are studying this problem. However, we can make some comments on the basis of our observations to date.

Firstly, we have published results [4, 5] on a 50-well QWSC which has a higher open-circuit voltage than the world's highest efficiency GaAs cell. The QWSC was grown by MBE at Philips Research Laboratories and processed as a 2.5 mm $\times$ 2.5 mm cell

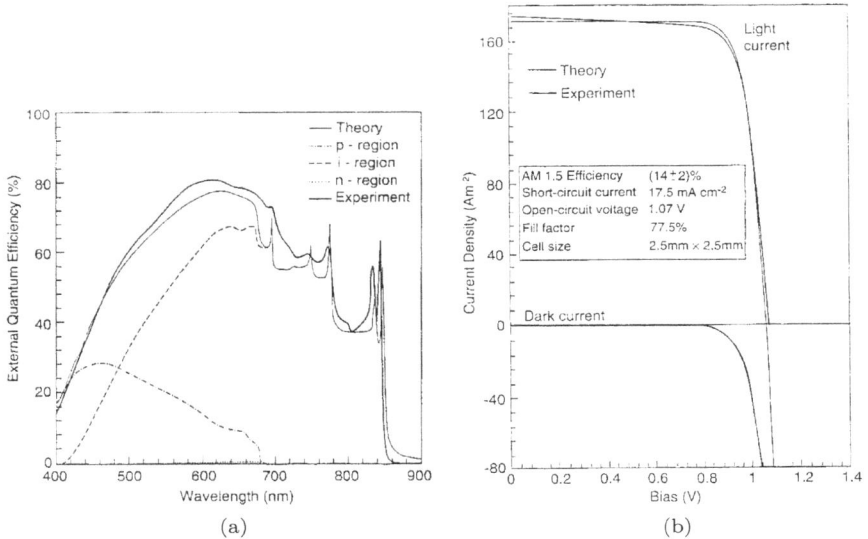

Figure 6.14:   (a) Results obtained on a 50 well MBE grown sample processed as a 2.5 mm × 2.5 mm solar cell. Experimental (heavy line) and calculated (fine line) spectral response at zero bias. Broken lines give calculated contributions from p-i-n regions. assuming an AM1.5 spectrum. (b) Results obtained on a 50 well MBE grown sample processed as a 2.5 mm × 2.5 mm solar cell. Current voltage characteristics under AM1.5 illumination in a solar simulator and in the dark. Full line is the sum of the fitted dark current and a short-circuit current calculated from the fit in (a) assuming an AM1.5 spectrum.

by collaborators at Eindhoven Technical University. The cell was measured to have 14% efficiency in air-mass AM1.5 conditions in a solar simulator at the ISE Calibration Laboratory, Freiburg. The spectral-response (external quantum efficiency at zero bias) and current-voltage characteristics of this cell in an AM1.5 spectrum are presented in Figs. 6.14(a) and 6.14(b).

It should be noted that the V_{oc} of this cell (1.07 V) is higher than that of the world's highest efficiency GaAs cell which is 1.022 V [12]. Hence in real practical material it is possible for a QWSC to produce a higher open-circuit voltage than the best currently available cell formed from the well material. As will be discussed in Section 6.5.7, we can see ways to enhance the short-circuit current of such AlGaAs/GaAs QWSCs. Note that as I_{sc} increases so will V_{oc}.

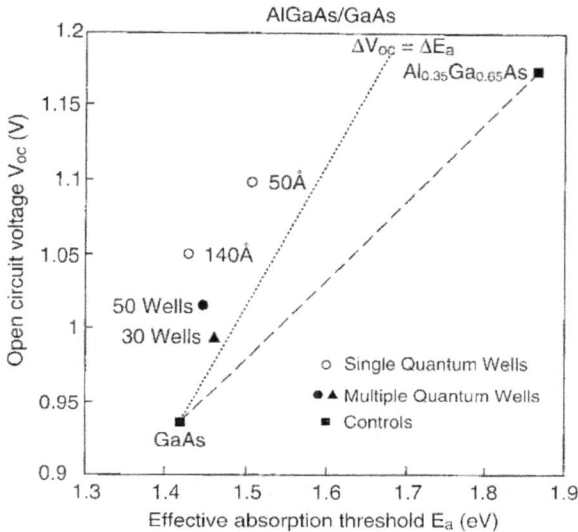

Figure 6.15: Open-circuit voltage (V_{oc}) versus effective bandgap for absorption (E_a) for GaAs and AlGaAs p-i-n control cells and for AlGaAs single-well and MQW cells with aluminum fraction around 35% under white light illumination (3000 K spectrum). The dotted line indicates the expected enhancement in V_{oc} above that of the GaAs cell due to confinement in the wells.

Secondly, QWSC dark currents are lower than expected. Experimentally we observe that QWSCs have dark currents which lie between the dark current of a homogeneous cell formed from the well material and the lower dark current of a conventional cell formed from the barrier material [4, 18]. However, the QWSC dark currents are one- to two-orders of magnitude lower (i.e., better) than a theoretical model (see [9, 19]) which assumes the quasi-equilibrium of the carriers in the barriers and wells (i.e., constant quasi-Fermi levels throughout the depletion region as assumed by Araujo and Marti).

Thirdly, as discussed further in [4, 18], the open-circuit voltage of QWSCs in real material does not follow the linear dependence on bandgap predicted for ideal material by Araujo and Marti. In Fig. 6.15, the variation of open-circuit voltage for some AlGaAs/GaAs cells and comparable AlGaAs and GaAs control cells plotted as a function of the effective bandgap for absorption (E_a).

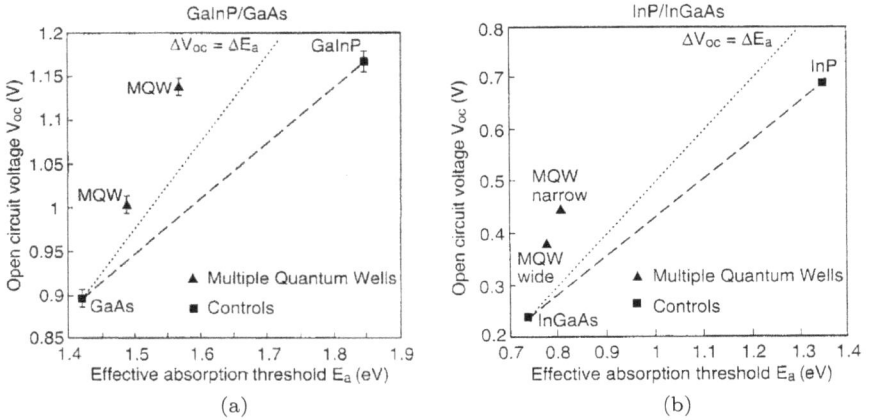

Figure 6.16: (a) The open-circuit voltage against effective bandgap for absorption under white light illumination as in Fig. 6.15: for quantum well cells in the GaAs and lattice matched GaInP system and comparable GaInP and GaAs control cells without wells. (b) The open-circuit voltage against effective bandgap for absorption under white light illumination as in Fig. 6.15: for quantum well cells in the InP and lattice matched InGaAs system and comparable InP and InGaAs control cells without wells.

In the case of the control cells E_a is the bulk bandgap. In the case of the QWSC it is the half-height point on the low-energy edge of the first excitonic feature in the spectral response of the well. It corresponds to the energy E_a in Fig. 6.10(a).

On adding GaAs quantum wells to an AlGaAs conventional cell the V_{oc} does not fall linearly with the effective bandgap for absorption as occurs in the model of Araujo and Marti in the radiative limit. Indeed we observe that the open-circuit voltages of AlGaAs/GaAs QWSCs can exceed that of a comparable GaAs cell by more than the shift in absorption edge.

We have recently observed this enhancement of V_{oc}, in QWSC in other materials. In Figs. 6.16(a) and 6.16(b) we show the V_{oc}, of lattice matched QWSCs in the GaInP/GaAs and InP/InGaAs systems, respectively.

In both cases it can be seen that the open-circuit voltage is enhanced over the comparable homogeneous cell formed from the well material (GaAs and InGaAs, respectively) by more than the increase expected from the change in the effective bandgap for absorption. Further details can be found in [18].

These experimental observations are in samples where non-radiative recombination dominates. However, we concluded from the dark-current modelling discussed in [19] that the recombination lifetime in the GaAs control cell was an order of magnitude longer than the AlGaAs control. Hence the GaAs is closer to ideal, radiative-limited material than the AlGaAs. On adding poor material to the GaAs in the form of AlGaAs barriers the voltage performance improves more than expected from confinement. It will be important to clarify how the quasi-Fermi level varies in real QWSC material, before concluding that the enhancement will not apply as the barrier materials become more ideal.

6.5.6   *Understanding the voltage dependence of QWSCs*

In order to understand the important new result that QWSCs enhance voltage above that expected from comparable control samples constructed of the material in the well, we have developed a program of theoretical and experimental studies to determine the quasi-Fermi level variation in QWSCs. We have started with the most straight-forward situation, namely a series of AlGaAs/GaAs single quantum well samples (SQWs). These are samples where we know from steady state photoconductivity and photoluminescence studies that the non-radiative contribution is always less than about 25% [6, 7].

Our approach is to study the photoluminescence (PL) signal due to the radiative recombination of carriers in the wells as a function of temperature and bias for a known generation rate of carriers from a laser which excites within the wells. As discussed in [9], the van Roosbroeck–Shockley equation, which is based on detailed balance arguments, relates the PL intensity to the absorption coefficient, which varies with photon energy and applied bias, and the difference between the quasi-Fermi levels of electrons and holes. Figure 6.17(a) shows that we can model the variation of the PL signal through the

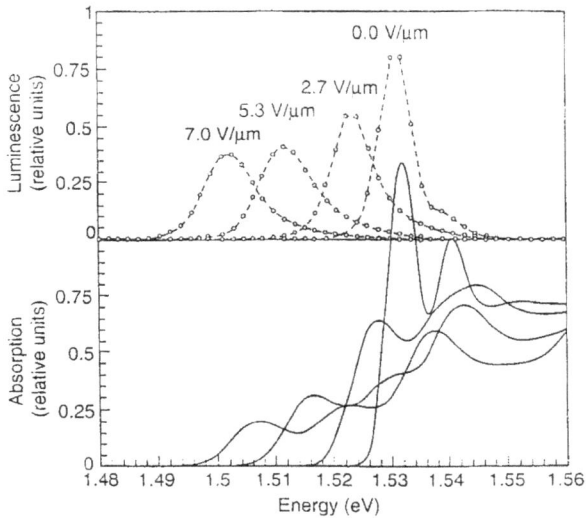

(a)

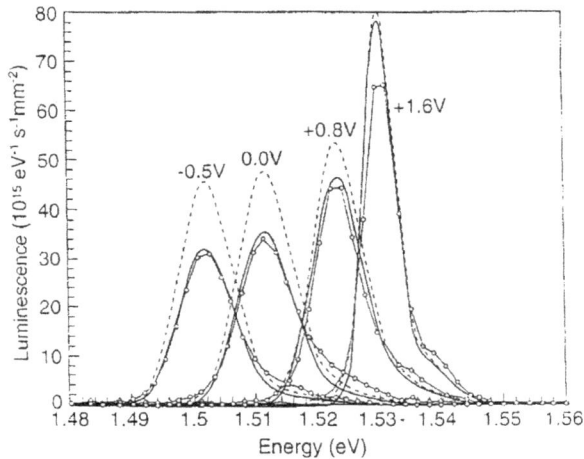

(b)

Figure 6.17: (a) Experimental photoluminescence intensity (upper diagram) measured at 40 K as a function of field across the single quantum well compared with the theoretical calculation (lower diagram) of the absorption in the quantum well. (b) Experimental luminescence intensity (circles) in absolute units at 40 K compared with the prediction of detailed balance (broken line) and following modification of the detailed balance prediction to allow for carrier escape from the well (full line).

effect of field on the QW absorption due to the quantum confined Stark effect.

The difference between our analysis and previous ones is that we make an absolute calibration of the PL intensity by determining the generation rate. This is done by measuring the saturation photocurrent at room temperature and in reverse bias where carrier escape dominates. We then equate this to the maximum integrated PL rate in forward bias, flat band conditions at low temperature where radiative recombination dominates. As we have a model for the absorption coefficient which predicts the magnitude and bias dependence in absolute units [9] we can compare the predictions of the van Roosbroeck–Shockley equation (broken line) with the measured PL intensity (circles) in absolute units, as shown in Fig. 6.17(b).

Though the theory predicts the PL peak position and width very well and the absolute level at flat-band (1.6V forward bias) it can be seen that the theory overestimates experiment at low forward and reverse biases. The full line is a fit to the experimental PL which indicates that carriers are lost by escape from the well and non-radiative recombination in the well, as determined by independent analyses of steady state photoconductivity and PL.

We conclude from this study that calculations of the limiting efficiency of QWSCs must include the escape of carriers to regions of higher bandgap. Further details on this work can be found in. [20, 21]. We are currently extending it to a comparison with the electro-luminescence which results from the radiative recombination of carriers injected into the wells under dark-current conditions.

6.5.7 *Optimizing the $Al_x Ga_{1-x} As/GaAs$ QWSC*

The $Al_x Ga_{1-x}As/GaAs$ material system is the best understood MQW system. As discussed in Section 6.5.6, we have produced a QWSC in this system with an open-circuit voltage greater than that of the world's highest efficiency GaAs solar cell. Hence the objective of our recent research on this system has been to enhance the short-circuit current.

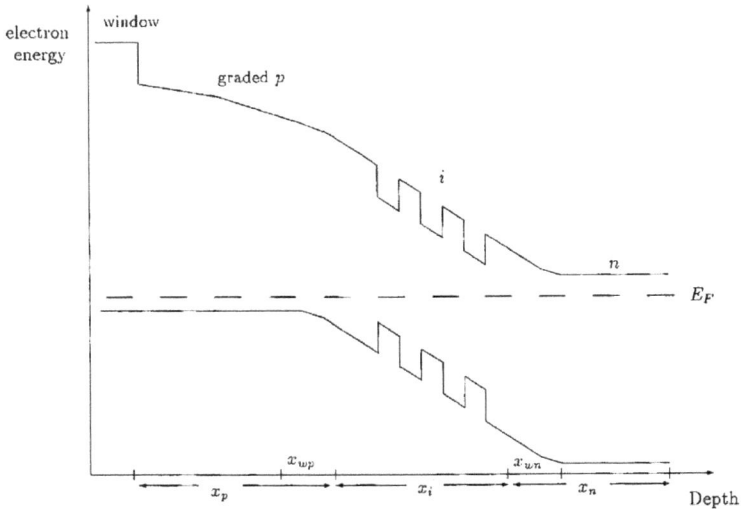

Figure 6.18:　Schematic diagram of a graded p-region quantum well solar cell as discussed in the text.

We have been studying two different approaches for enhancing the current output for light of energy above and below the AlGaAs bandgap (E_g in Fig. 6.10(a)). For short wavelength, high energy light the main problem is the inherently poorer AlGaAs material quality than GaAs. Hence we have investigated graded bandgap approaches in which the aluminum content increases in the p-region towards the top surface as shown in Fig. 6.18. The resulting bandgap increase pushes minority carrier electrons towards the field-bearing i-region. Furthermore a higher proportion of the incident light reaches the i-region where it is more efficiently converted into current. We have developed a computer model for carrier generation and collection in such a cell and this reproduces the measured spectral response well as shown in Fig. 6.19(a).

For long wavelength, low energy light, the problem is that we can only incorporate up to 50 quantum wells which absorb only around 50% of the light. Background doping in the i-region needs to be low if the built-in field is to be maintained across the wells at the operating bias [3]. Such background dopings ($\sim$ few $\times$ 10^{14} cm^{-3}) have been

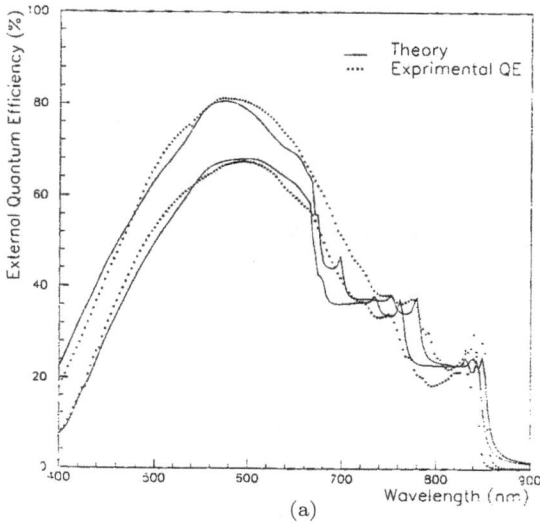

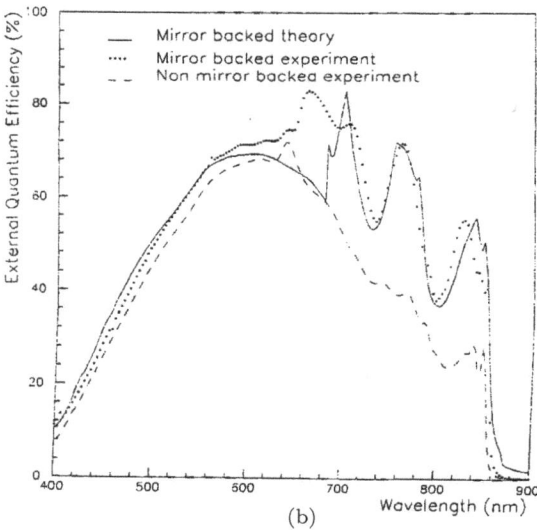

Figure 6.19: (a) The measured spectral response (external quantum efficiency) as a function of wavelength at zero bias compared with the theoretical models described in [22] for: two 30 well AlGaAs QWSCs one with a graded bandgap in the p-region as in Fig. 6.18 (upper pair) and the other (lower pair) with a constant p-region bandgap. (b) As in (a) but for two 30 well AlGaAs QWSCs one with a mirror at the back surface to double the passage of the light through the wells.

achieved in the sample presented in Fig. 6.14, but in general they are $\sim$ few $\times$ 10^{15} cm^{-3} in the AlGaAs system, which precludes the addition of more than around 50 wells. We have therefore investigated now the addition of a rear-mirror, which reflects the light back through the wells, can improve light absorption and hence current output. Our collaborator, Malcolm Pate, at the EPSRC III–V Facility at the University of Sheffield has successfully added back-surface mirrors to both the 1 mm circular photodiodes we use as test devices and to our 3 mm $\times$ 3 mm solar cells. We have extended our spectral response model to incorporate this modification. It reproduces the enhancement observed in a 30 quantum well sample as seen from Fig. 6.19(b).

We are currently attempting the growth of a sample which incorporates both the graded bandgap and 50 wells. The prediction of this model for the spectral response with both improvements is compared with the experimental results for our 14% cell in Fig. 6.20.

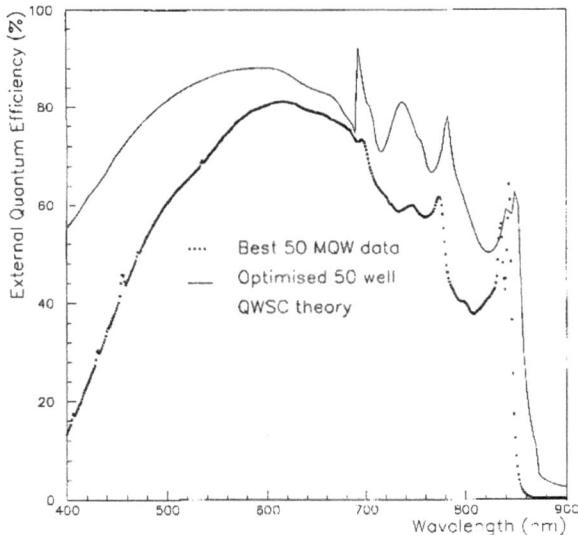

Figure 6.20: Prediction of the models used in Figs. 6.19(a) and 6.19(b) for the spectral response of a 50 well QWSC with both the improvements in Figs. 6.19(a) and 6.19(b), compared with the measured spectral response of the 14% efficient 50 well QWSC in Fig.6.14.

We predict that a cell with both improvements would have an AM1.5 efficiency of 22% assuming the fill-factor remains the same as in Fig. 6.14. Further details are given in [22].

6.5.8 *GaAs/InGaAs strained QWSCs*

GaAs solar cells hold many of the efficiency records for conventional solar cells [12]. Hence, adding quantum wells to a high efficiency GaAs cell should be particularly advantageous. However, there is no suitable semi-conductor with a smaller bandgap and the same lattice constant as GaAs. The InGaAs wells which are extensively studied for lasers have a wider lattice spacing and so the quantum well material is strained, leading to a number of growth problems.

We, and other groups who have studied QWSCs in this system, have established [13–15] that, though the addition of InGaAs wells enhances the short-circuit current of GaAs cells, this is presently not sufficient to overcome the voltage loss which is very sensitive to material quality.

We therefore initiated a systematic study of the relationship between the strain in the quantum wells and the forward dark currents (which determine the voltage behavior) in GaAs/InGaAs multiquantum well systems. Such an investigation had not previously been made at the low forward biases where solar cells operate. We studied the dark-current of a series of different MQW geometries varying well and barrier widths, well depth (indium fraction), number of wells and i-region thickness but keeping the GaAs p-region top layer similar in all samples. We estimated the number of misfit dislocations in the material by measuring the recombination dark-line densities observed in cathodoluminescence (CL) in collaboration with Max Mazzer and co-workers in the Materials Department, Imperial College. We studied a range of samples grown by both MBE and MOVPE.

A strong correlation between the dark-current density at fixed bias and the degree of strain relaxation determined by the dark-line density measured at the top interface (i.e., between the MQW and the p-region) was observed, as shown in Fig. 6.21.

The study showed that a significant parameter was the excess strain before relaxation for each sample defined by the difference

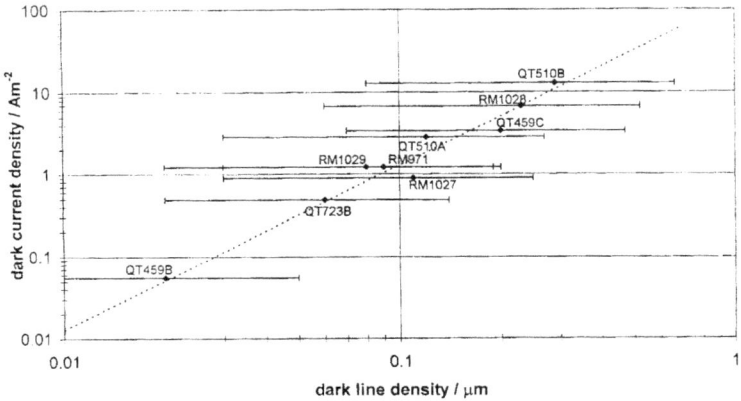

Figure 6.21: Dependence of the dark current density of GaAs/InGaAs strained QWSCs at a fixed bias of +0.5V on the dislocation density as determined by the dark line density observed in cathodoluminescence.

between the average strain (calculated from the indium fraction, number of wells, width of wells and barrier thickness) and the equilibrium strain according to the theory of Mathews and Blakeslee [23].

When either the dark-line density or dark current density are plotted against excess strain as in Figs. 6.22(a) and 6.22(b), it is observed that there are levels of excess strain where both the dark current and the dark-line density suddenly rise.

The first increase we identify with the MQW region relaxing causing many misfit dislocations at the bottom interface (i.e., between the MQW and the n-region) and the second increase can be explained by the formation of extra misfit dislocations at the top interface between the p-region and the MQW.

Misfit dislocations below the wells have been observed by high-resolution transmission electron microscopy in cross sections of these samples (Fig. 6.23).

The misfit dislocations are situated in the n-region at the interface with the bottom QW and they relax the strain locally. They lie on (111) planes and appear dissociated into two partial dislocations with a stacking fault between them. By observing a large number of adjacent areas in this sample (QT510A) we counted an average density of 1.7 misfit dislocations per micron. We did not observe any

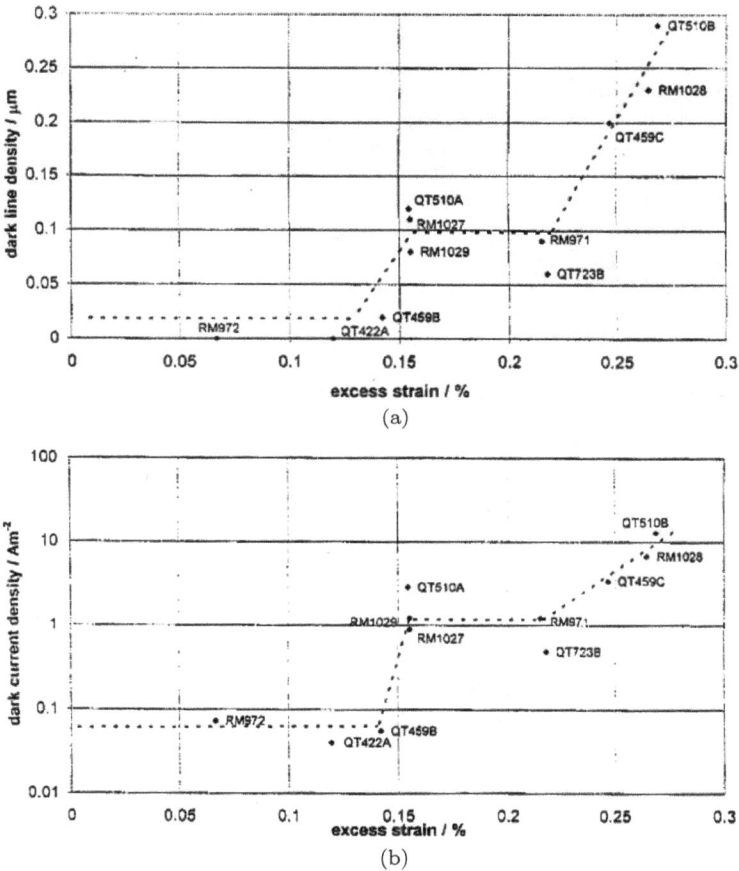

Figure 6.22: (a) The *excess strain* before relaxation as defined in the text in terms of the equilibrium strain according to Mathews and Blakeslee, for GaAs/InGaAs strained MQW samples compared to the dislocation density given by the dark-line density in cathodoluminescence. (b) The *excess strain* before relaxation as defined in the text in terms of the equilibrium strain according to Mathews and Blakeslee, for GaAs/InGaAs strained MQW samples compared to the dark current density at +0.5 V.

misfit dislocations in sample QT422A (Figs. 6.12 and 6.22). Part of this work has been submitted for publication [24].

This study has provided the information required to optimize a QWSC in this system. A cell has been designed with excess strain such that the dark current is on the lowest plateau while having the

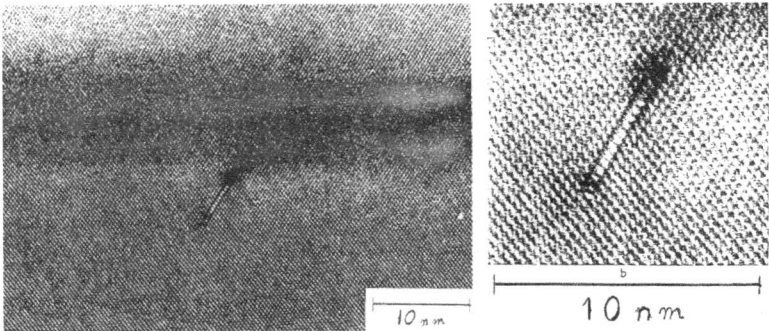

Figure 6.23:   High-resolution transmission electron micrograph of a (110) cross-section of GaAs(001)/InGaAs MQW (QT510A) showing: (left) Misfit dislocation at the interface between the bottom QW (InGaAs — dark band) and the n-region (GaAs — clear region). The white dots in both areas represent the atomic positions; lines of white dots correspond to the projection of atomic planes; (right) Magnified micrograph of the area containing the misfit dislocation.

maximum number of wells to provide the maximum current gain. Our calculations suggest that even in this optimized cell it will be necessary for the light to make more than one pass through the wells to obtain sufficient I_{sc} enhancement to compensate for the voltage loss. We have therefore extended the back-surface mirror technique discussed in the previous section to the GaAs/InGaAs system. This required growth of an AlGaAs etch stop to enable the substrate to be removed. The highly-doped GaAs substrate absorbs rather strongly just below GaAs bandgap. The mirror produces significant improvement in the spectral response as shown in Fig. 6.24.

It can be seen that the mirror is considerably more effective than a Bragg reflector. The optimized cell has recently been grown at Sheffield and is currently being processed.

6.5.9   *InP/InGaAs lattice matched QWSCs*

The InP bandgap is well matched to the solar spectrum and the InGaAs wells are lattice matched and relatively deep so they should give a very significant efficiency enhancement. InP cells are also favored for space applications because of their very good radiation

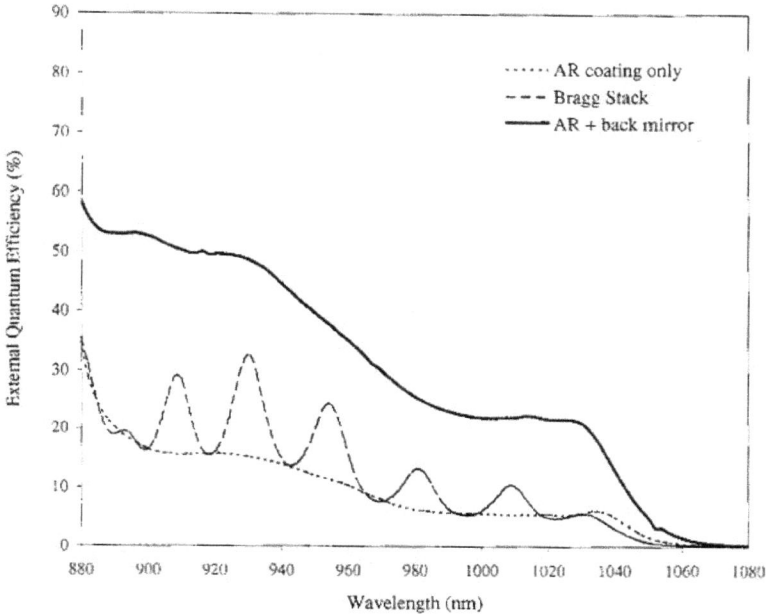

Figure 6.24: The spectral response (external quantum efficiency) as a function of wavelength at zero bias for a GaAs QWSC with 10 strained wells and two similar samples with 10 either a Bragg-Stack reflector or a rear-surface mirror to double the passage of the light through the wells.

tolerance. However, we have found that, though the wells are not strained, they have another material problem. Both InP and InGaAs can be grown individually with acceptable background doping, but InP/InGaAs MQWs have high i-region background doping values. This occurs when samples are grown by MOVPE or MBE at Sheffield, by MOVPE in an industrial laboratory and at the Paul Scherrer Institute (PSI) Zurich.

The problem appears to result, primarily, from diffusion of the p-dopant into the i-region. It is possible that the diffusion is enhanced by the presence of imperfections in the wells. We therefore decided, in collaboration with John Epler at PSI, to grow cells in which the InP p-region was not intentionally doped. We relied on the diffusion of the p-dopant, which was zinc in this case, from a highly doped

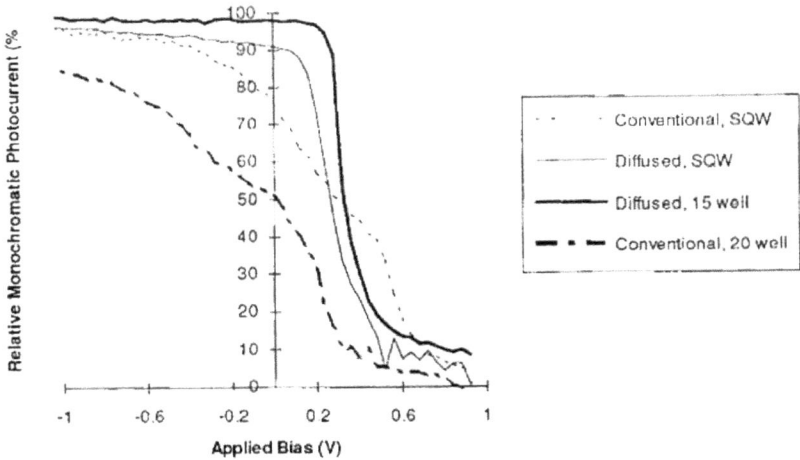

Figure 6.25: The relative monochromatic photocurrent (RMP) at 900 nm wavelength as a function of applied bias for InP/InGaAs single and MQW samples: full lines — with diffused p-doping; broken lines — with conventional doping.

InGaAs cap into the undoped InP i-region buffer layer to produce the p-region emitter. This novel approach improved the background doping very significantly.

We have two main methods for assessing the background doping levels. Firstly we measure photocurrent at fixed wavelength as a function of bias, a measurement which we call relative monochromatic photocurrent (RMP).

This measurement has proved to be particularly sensitive to the i-region residual background doping of $N_i \sim 10^{15}\,\mathrm{cm}^{-3}$ which our studies of the AlGaAs/GaAs system [3] showed was the upper limit for the built-in field to be maintained across the i-region in forward bias. For higher background dopings the collection efficiency and the RMP fall at biases below that at which the dark-current cancellation takes over, as shown for the conventional single and MQW samples in Fig. 6.25. The novel doping approach improves the RMP dramatically. For both single and MQW diffusion doped samples it remains at its reverse bias value well into forward bias as shown in Fig. 6.25.

The second, and more quantitative, method of investigating background doping levels of this magnitude is to fit the measured spectral response allowing for the effect of N_i. In Figs. 6.26(a) and 6.26(b) we show the experimentally determined spectral response for comparable conventionally doped and diffused p-layer samples, respectively.

The theoretical model allows for the reduction in transport efficiency if the field only extends over part of the i-region due to the level of the background doping. The good fit to both sets of spectral response data, particularly on the first sub-band continuum in the quantum well, is obtained primarily by changing the N_i which is 2×10^{16} cm^{-3} in the conventional case and less than 10^{15} cm^{-3} in the diffused p-layer case.

These values agree with capacitance–voltage measurements though the latter are not easy to make in the presence of highly doped n and p regions.

We have compared the performance of the 15 well diffused p-layer sample in our 3000 K white-light source with two control cells. One is an InP p-i-n homostructure. The other has an i-region formed from lattice matched InGaAs and so the sample is a double heterostructure. In both cases the InP p-region was doped by diffusion from an InGaAs cap in the same manner as for the QWSC. The results obtained in our white light source are shown in Fig. 6.27. It can be seen that the I_{sc} of the QWSC is higher than that of the InP (barrier) control and the V_{oc} is higher than the InGaAs (well) control. These results will be presented in the following section after correction to AM1.5 conditions. It should be noted that the corrected efficiency of the QWSC is higher than the InGaAs control. This is the first time that a QWSC has been shown to have higher efficiency than a comparable control made from the well material.

6.5.10   *Temperature dependence of QWSC efficiency*

As mentioned in Section 6.5.2, QWSCs are particularly suited to the light concentration systems which bring the costs of solar electricity down. Solar cells get hot under concentrated light and lose efficiency as the bandgap shrinks with increasing temperature. The barrier

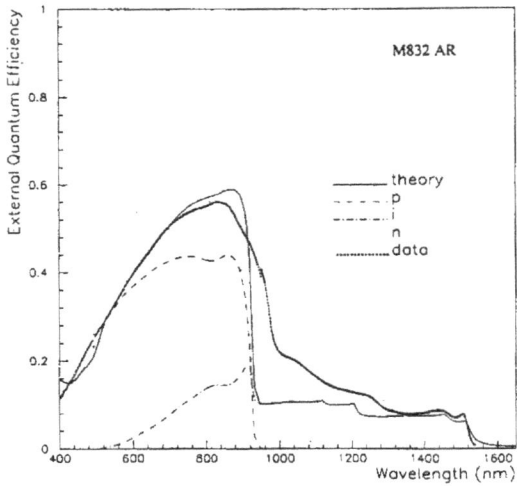

(a)

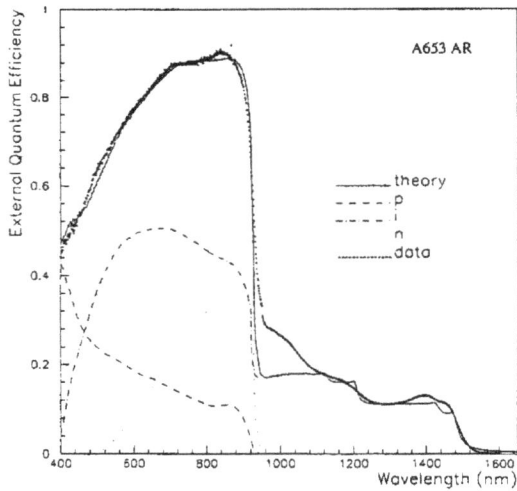

(b)

Figure 6.26: (a) Heavy line is experimentally determined spectral response compared to the theoretical fit discussed in the text for a conventionally doped 20 well sample with high background doping (2×10^{16} cm^{-3}). (b) Heavy line is experimentally determined spectral response compared to the theoretical fit discussed in the text for a diffused p-layer 15 well sample with low background doping ($> 10^{15}$ cm^{-3}).

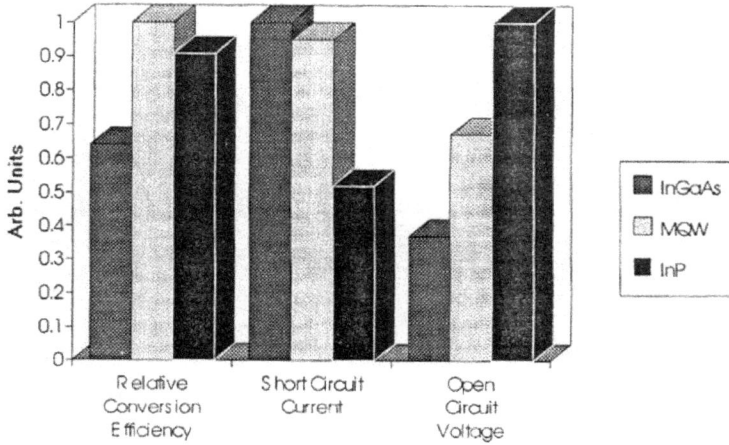

Figure 6.27: Results on the relative conversion efficiency, relative short-circuit current and relative open-circuit voltage measured in a 3000 K white light spectrum for three comparable samples all with diffused doping of an InP emitter. The 15 well MQW sample is compared with two control samples, one with an i-region of similar width containing the barrier material InP and the other a double heterostructure sample with an i-region containing the well material InGaAs.

and well bandgaps also shrink in a QWSC. However, our studies of carrier escape from single quantum wells [6] have demonstrated the important contribution thermal activation plays in carrier escape. In deep wells, such as in the InP/InGaAs system discussed in the previous section, carrier escape may be less than 100% efficient at room temperature. Then, as temperature rises, one would anticipate that the well contribution would increase leading to a more favorable temperature coefficient of efficiency.

We have designed a Peltier heating system to control the temperature of our samples and have studied how the efficiency and other cell parameters vary with temperature. Preliminary results on the temperature dependence of the efficiency, short-circuit current, open-circuit voltage and fill-factor of the diffused p-region InP/InGaAs QWSC and control samples discussed in Section 6.5.10, are presented in Fig. 6.28. The measurements were all performed in a 3000 K white-light source. The room temperature efficiencies, I_{sc}, V_{oc}

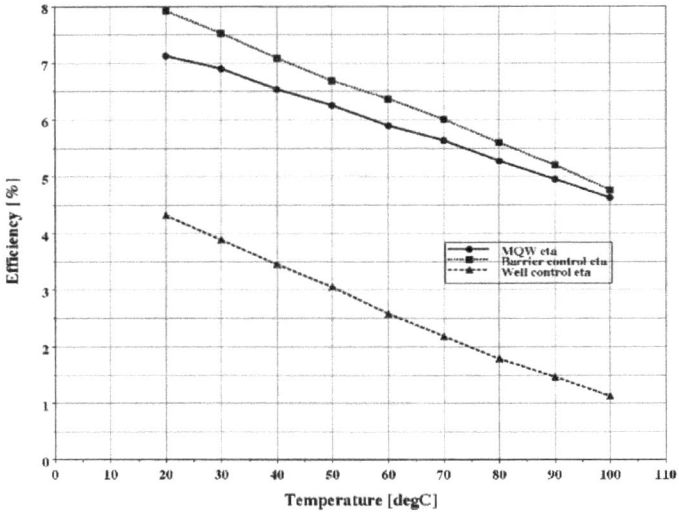

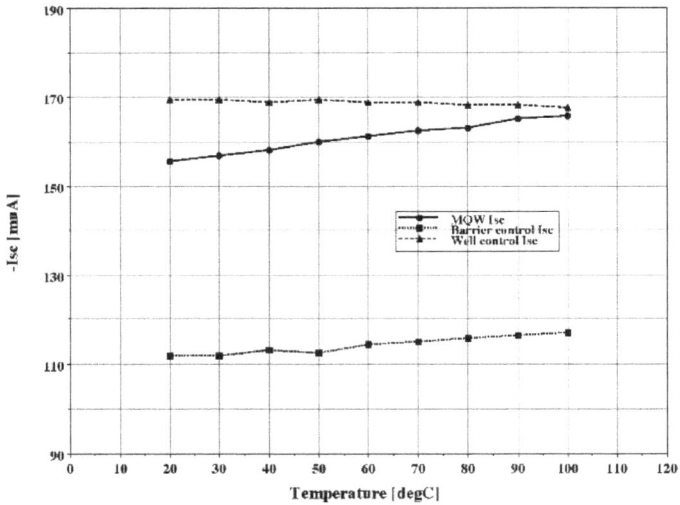

Figure 6.28:   The temperature dependence of the efficiency, short-circuit current, open-circuit voltage and fill factor of the InP/InGaAs QWSC and the two control cells presented in Fig. 6.27. The measurements were all made in the 3000 K light source as in Fig. 6.27. The data have been scaled by correcting the room temperature values to those appropriate to an AM1.5 spectrum using the measured spectral response.

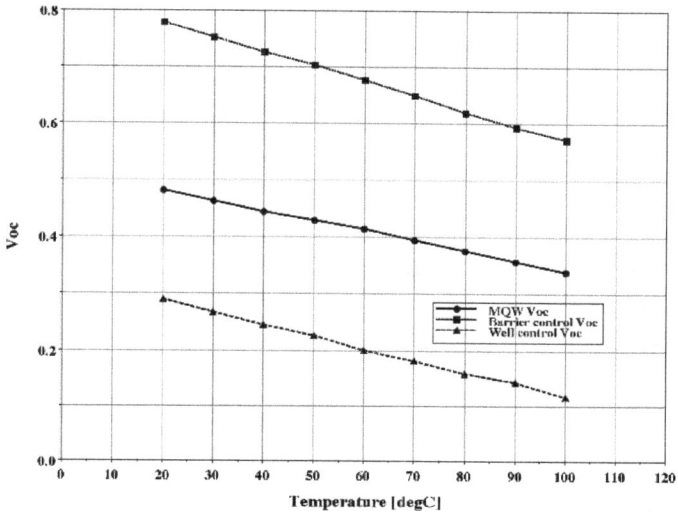

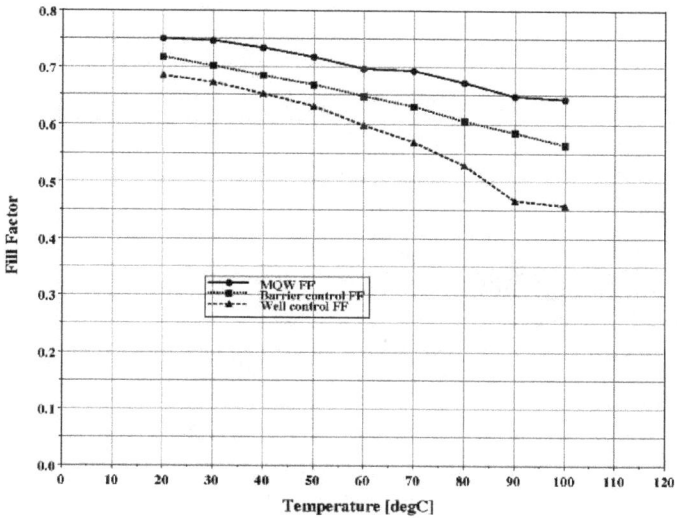

Figure 6.28: (*Continued*)

and fill-factor values were corrected to AM1.5 conditions using the measured spectral response of the devices.

It can be seen that the efficiency of the QWSC falls less steeply with temperature than either the InP homogeneous control or the InGaAs heterostructure. The measured temperature coefficients at 25°C are for the QWSC (-32×10^{-3} %/K), for the InP control (-40×10^{-3} %/K) and for the InGaAs control (-41×10^{-3}%/K).

In the AlGaAs system the QWSCs also have better temperature coefficients of efficiency than either barrier or well control cells and the effect gets more marked as the aluminum composition increases and the wells get deeper [25].

6.5.11   *Conclusions*

It has now been clearly established experimentally by a number of groups in a number of material systems, that the QWSC can enhance I_{sc} over a comparable conventional cell formed from the barrier material and that AlGaAs/GaAs QWSCs enhance efficiency over comparable AlGaAs cells. We have recently made the important observation, in three material systems, that the V_{oc} is enhanced over that of a comparable conventional cell formed from the well material by more than the shift in the absorption edge. Although these observations are made in material in which non-radiative recombination dominates, we have argued that it will be important to establish exactly how the quasi-Fermi levels vary in a QWSC before concluding that this will not happen should ideal, radiative dominated material ever be produced. Our determination of the quasi-Fermi level separation in SQW samples is a first step in this direction. By determining the photoluminescence intensity in absolute units we have demonstrated that the escape of carriers to wider bandgap regions must be taken into account in any detailed balance arguments. We have demonstrated a 50 well QWSC with a higher open-circuit voltage than the world record cell constructed from the well material. Furthermore we have shown theoretically how the short-circuit current can be enhanced by grading the bandgap and adding back-surface mirrors. We have demonstrated both improvements separately. In the GaAs/InGaAs system we have established a strong correlation between the dark-line density and the dark current. We have established the importance of excess strain and demonstrated two separate plateaus in the dark current versus excess strain plot. We have shown that it may be possible to enhance efficiency in this system with the use of a back mirror. In the InP/InGaAs system we have demonstrated a novel method of p-region doping that improves the background doping level in the i-region. In this way we have demonstrated for the first time that a QWSC can have higher efficiency than a comparable control heterostructure cell incorporating the well material. In these cases the QWSC has superior efficiency variation with temperature than control cells made from barrier or well material.

Finally we believe that, as the nature of the variation of the quasi-Fermi levels in the QWSC become clearer, quantum well engineering may be employed to minimize both radiative and non-radiative recombination. In fact, in the original proposal [2], it was envisaged that the simple QWSC of Fig. 6.10 would be the first of a number of more complex solar cells which would exploit the many and varied structures which have now been made available for other optoelectronic applications by modem crystal growth techniques.

6.5.12 *Acknowledgements*

We are grateful to Axel Schönecker, Klaus Schitterer and Michaela Burs (ISE Calibration Laboratory) for use of the solar simulator. We wish to thank Peter Nouwens, Peter Ragay and Professor J.H. Wolter (Eindhoven University of Technology) for processing help, and Geoff Duggan (Sharp Laboratories of Europe) and Craig McFarlane (UNL) for theoretical assistance. We have not had space in this article to discuss the time-resolved measurements of our samples which have been made by Chris Phillips and George Thucydides (Imperial College) and by Phil Dawson and Andrew Heath (UMIST) but we wish to acknowledge the importance of such studies. We are extremely grateful to the growers who have provided us with samples, in particular Tin Cheng and Tom Foxon (Nottingham), John Epler (PSI, Zurich), Baldev Dosanjh and Christine Roberts (Semiconductor IRC), Chris Button, Bob Grey, Mark Hopkinson and

John Roberts (Sheffield). The processing provided by Geoff Hill and Malcolm Pate (Sheffield) has been particularly important. We are most grateful to Max Mazzer and Carlo Zanotti-Fregonara for their EBIC and CL characterization of our samples and for advice on the material quality of strained systems. We would also like to thank Professor Tony Stradling (Imperial College), Professor Bruce Joyce and Professor Gareth Parry (Semiconductor IRC) for their support and the use of facilities. We wish to acknowledge financial support from the Central Research Fund of the University of London, Clean Technology Unit, EU Human Capital and Mobility program, EPSRC, Greenpeace Trust, Mark Leonard Trust and Tata Ltd (India).

6.5.13 *References*

[1]   C.H. Henry, *J.Appl.Phys.* 51, 1980, 4494.

[2]   K.W.J. Barnham and G. Duggan, *J.Appl.Phys.* 67, 1990, 3490.

[3]   K.W.J. Barnham, B. Braun, J. Nelson, M. Paxman, C. Button, J.S. Roberts, and C.T. Foxon, *Appl. Phys. Lett.* 59, 1991, 135.

[4]   K. Barnham, T.Ali, J. Barnes, J. Connolly, G. Haarpaintner, J. Nelson, J. Osborne, E. Tsui, A. Zachariou, C. Button, M. Pate, J. Roberts, and T.Foxon, *Optoelectron. — Dev.Technol.* 9, 1994, 483.

[5]   K. Barnham, J. Barnes, G. Haarpaintner, J. Nelson, M. Paxman, T. Foxon, and J. Roberts, *MRS Bulletin*, October 1993, 51.

[6]   J. Nelson, M. Paxman, K.W.J. Barnham, J.S. Roberts, and C. Button, *IEEE J. Quantum. Electron.* 29, 1993, 1467.

[7]   J. Barnes, E.S.M. Tsui, K.W.J. Barnhart, S.C. Macfarlane, C. Button, and J.S. Roberts, "Steady state photocurrent and photoluminescence from single quantum wells as a function of temperature and bias", *J.Appl. Phys.* 1996.

[8]   G. Thucydides, J.M. Barnes, E. Tsui, K.W.J. Barnharn, C.C. Phillips, T.S. Cheng, and C.T. Foxon, *Semicond. Sci. Technol.* 11, 1996, 331.

[9]   J. Nelson in M.H. Francome and J.L. Vossen (eds.), *Physics of Thin Films*, Vol. 21, Academic Press, 1995, p. 311.

[10] F.W. Ragay, J.H. Wolter, A. Marti, and G.L. Araujo, *Proceedigns of the 12th European Photovoltaic Solar Energy Conference*, Amsterdam, H.S. Stephens and Associates, 1994, p. 1429.

[11] G.-L. Araujo and A. Marti, *Proceedigns of the 11th E.C. Photovoltaic Solar Energy Conference*, Harwood Academic Publishers, 1992, p. 142. and G.L. Araujo, A. Marti, F.W. Ragay, and J.H. Wolter, *Proc. 12th European Photovoltaic Solar Energy Conference*, Amsterdam, H.S. Stephens and Associates, 1994, p. 1481.

[12] M.n A. Green, K. Emery, K. Bucher, and D.L. King, *Prog. Photovolt.: Res. Appl.* 3, 1995, 229.

[13] J. Barnes, T. Ali, K.W.J. Barnham, J. Nelson, E.S.M. Tsui, J.S. Roberts, M.A. Pate, and S.S. Dosanjh, *Proceedigns of the 12th European Photovoltaic Solar Energy Conference*, Amsterdam, H.S. Stephens and Associates, 1994, p. 1374.

[14] F.W. Ragay, J.H. Wolter, A. Marti, and G.L. Araujo, *Proceedigns of the 1st World Conference on Photovoltaic Energy Conversion*, Hawaii, IEEE, 1994, p. 1754.

[15] Y. Yazawa, T. Kitatani, J. Minemura, K. Tamura, and T. Warabisako, *Proceedigns of the 1st World Conference on Photovoltaic Energy Conversion*, Hawaii, IEEE, 1994, p. 1878.

[16] R. Corkish and M. Green, *Proceedigns of the 23rd IEEE Photovoltaic Specialists Conference*, IEEE, 1993, p. 675.

[17] N.G. Anderson, *J. Appl. Phys.* 78, 1995, 1850.

[18] K. Barnham, J. Connolly, P. Griffin, G. Haarpaintner, J. Nelson, E. Tsui, A. Zachariou, J.Osborne, C. Button, G. Hill, M. Hopkinson, M. Pate, J. Roberts, and T. Foxon, "Voltage enhancement in quantum well solar cells", *J. Appl. Phys.* 80, 1996, 1201.

[19] J. Nelson, K. Barnham, J. Connolly, G. Haarpaintner, C. Button, and J. Roberts, *Proceedigns of the 12th European Photovoltaic Solar Energy Conference*, Amsterdam, H.S. Stephens and Assoc., 1994, p. 1370.

[20] J. Nelson, B. Kluftinger, E.S.M. Tsui, K. Barnham, J.S. Roberts, and C. Button, "Quasi-fermi level separation in quantum well solar cells", *Proceedings of the 13th European Photovoltaic Solar Energy Conference*, Nice, October 1995.

[21] E.S.M. Tsui, J. Nelson, and K. Barnham, "Determination of the quasi-Fermi level separation in single quantum well p-i-n diodes", *J.Appl.Phys.* 80, 1996.

[22] J. Connolly, K.W.J. Barnham, and C. Roberts, *Proceedings of the ISES World Solar Congress*, Harare, 1995.

[23] J.W. Mathews and A.E. Blakeslee, *J.Cryst. Growth* 27, 1974, 118.

[24] P.R. Griffin, J. Barnes, K.W.J. Barnham, M. Mazzer, C. Zanotti-Fregonara, C. Olson, C. Rohr, G. Haarpaintner, J.P.R. David, J.S. Roberts, R. Grey, and M.A. Pate, "Effect of strain relaxation on forward bias dark currents in GaAs/InGaAs multi-quantum well p-i-n diodes" *J. Appl. Phys.* 80, 1996, 5815.

[25] I. Ballard, "Temperature effects on the efficiency of multi-quantum well solar cells", M.Sc. Thesis, Imperial College of Science Technology and Medicine, London, 1995.

6.6 Discussion following Keith Barnham's presentation

Emery: Your devices seem to be very small and that indicates to me a high perimeter recombination component. What makes you think that this is not causing problems with your dark current and other effects such as defects?

Barnham: We probably are losing some V_{oc} because of edge effects. However, our first photodiode devices were 1 mm × 1 mm in size and we then "scaled up" to 3 mm × 3 mm and did not find any noticeable improvement in dark current. So, either these perimeter effects are not very important or, we will only eliminate them by going to very much larger cells.

Goren: Your quantum well structure leads to an abrupt change of the periodic potential. I would guess that this might lead to the formation of recombination centers in the forbidden gap at the interface. Do you have an estimate of the recombination velocity at this interface?

Barnham: I cannot give you a quantitative answer but there are several reasons to believe that it is negligibly small.

Firstly, we have made extensive studies of the temperature-dependence and bias-dependence of the carrier escape efficiency. In particular, we look at carrier escape by measuring the photocurrent and photoluminescence measurements give us the radiative recombination rate. All of these measurements fit a simple model that does not require any non-radiative or surface effects. You can study the details in our published papers but if pressed I'd estimate that such effects could be at most 20%, for essentially all of the carriers escape before recombination — even in our average-quality AlGaS quantum wells.

Secondly, do not forget that quantum wells are a highly developed technology in the laser field. There it is well known that the radiative recombination efficiency of a quantum well is considerably higher than it is for a double heterojunction structure. I can see no reason to expect a different situation to hold with solar cells.

Rosenwaks: I would imagine that one of your main problems is the diffusion length in the i-, or multi quantum well region. Do you have an estimate for this diffusion length?

Barnham: This isn't a problem. In the i-region you have the drift field which is easy to maintain due to low background doping. This results in a very efficient i-region.

Rosenwaks: Have you thought of going to a super lattice with its associated mini band transport? This could overcome your present problem of having to get the carriers to travel over or tunnel through the barriers in your multi quantum well.

Barnham: I think that this is a good idea. We actually thought of this possibility at the start of the research but have never worked on it. We are still experiencing materials problems in maintaining the field and super lattices may provide a solution. We'd be delighted to collaborate with anyone else who'd like to work on this aspect.

Emery: Is a quantum well cell better than a tandem cell of the same material and if so, what makes it better?

Barnham: Tandem cells have been around for much longer and have already produced some truly impressive results. In particular, they will probably always be able to achieve higher voltages than our cells because you can simply stack more components in tandem. However, they have an ohmic resistance problem which we do not. As a result, we will always be able to achieve higher currents, particularly since there is such a high efficiency for getting the carriers out of our wells. This advantage, together with our low temperature coefficient should really become apparent under high concentration illumination.

6.7 Addendum to Prof. Barnham's presentation (May 2020)

My 1995 presentation reported the observation of a new phenomenon in these cells: suppression of the radiative recombination from the QW. The effect was parametrized as a reduction in the quasi-Fermi level separation (QFLS) between electrons and holes in the QW below the output voltage of the cell.

The publication of these results led to considerable controversy. Theorists claimed that in radiatively dominated solar cell the QFLS had to equal the output voltage at all points. They argued that the efficiency of a radiatively dominated QW cell could not exceed the efficiency of an ideal bulk cell with the same absorption edge. The theoretical controversy abated in 2003 when our group pointed out [1] that if the carriers in the QW were recombining at a temperature higher than the carriers in the bulk, there would be a thermo-electric force driving current from the wells, thus reducing the number of carriers recombining in the well and the QFLS. The presence of hot carriers in 1, 5 and 10 strain-balanced QWs at biases corresponding to around 200x concentration was established by fitting carrier temperatures and QFLS in electro-luminescence spectra [2].

In 2012, a 42.5% triple-junction cell with QWs in the top and middle cell was demonstrated [3]. The middle QW cell was radiatively dominated and was responsible for the significantly higher efficiencies and output voltages achieved compared to an identical triple junction cell without QWs. These results are consistent with raised carrier recombination temperatures and reduced QFLS in the middle cell.

Seventeen years after suppressed radiative recombination in QW solar cells was first reported at the 1995 Symposium, the material quality of commercial concentrator cells had improved to the extent that the recombination in the QWs of the middle component of a triple junction cell was radiatively dominated. Efficiency and voltage enhancement were observed, a QW effect that some theorists had predicted was impossible assuming cell and well were in thermal equilibrium.

6.7.1 *References*

[1] M. Mazzer, K.W.J. Barnham, N. Ekins-Daukes, D.B. Bushnell, J. Connolly, G. Torsello, D. Diso, S. Tundo, M. Lomascolo, and A. Licciulli, "The use of III–V quantum heterostructures in PV: thermodynamic issues", *Proceedings of the 3rd World Conference on PV Energy Conversion, WCPEC3*, 2003, p. 2661.
[2] M.F. Führer, Ph.D. Thesis, Imperial College London, 2010.
[3] B.C. Browne, J. Lacey, T.N.D. Tibbits, G. Bacchin, T.C. Wu, J.Q. Liu, X. Chen, V. Rees, J. Tsai, and Jan.-G. Werthen "Triple-junction quantum-well solar cells in commercial production", *Proceedings of the 9th International Conference Concentrator PV Systems, CPV-9*, 2013.

Chapter 7

1997

7.1 Editor's Foreword

In Chapter 5, which featured the November 1994 keynote lectures, the late Steven Kaneff presented details of a utility-scale concentrator photovoltaic system based upon a giant parabolic dish he had developed in Australia. In the following symposium, which took place in March 1996, we had hoped to be able to discuss two other utility-scale concentrator PV technologies: One, the European *EUCLIDES* project, which availed itself of the linear parabolic trough concept (that had proved itself so successful for solar-thermal systems: see volume 1 of this history), was to have been presented by Antonio Luque. The other, the Amonix Fresnel-lens system, was to have been presented by that company's founder, Vahan Garboushian. Unfortunately, both speakers were prevented from coming to Sede Boqer each for reasons that need not concern us here. However, we were fortunately able to re-schedule them for the 8th Symposium, which took place in November 1997. Antonio was able to update his presentation accordingly, but Vahan had, in the meantime, published his planned presentation in *Solar Energy and Solar Cells* 47 (1997) 315–323.

7.2 Photovoltaic concentration (Prof. Antonio Luque)

A Keynote lecture presented by Professor Antonio Luque (Instituto de Energia Solar, Universidad Politecnica de Madrid, Spain)

7.2.1   *Introduction*

At the present time solar cells are too expensive to make photovoltaics cost effective. The main motivation of concentration is to reduce the amount of these expensive solar cells by employing low-cost light collecting surfaces. In concentration photovoltaics such optical surfaces are, almost always, mirrors or lenses. These optical elements deflect the sun's rays and focus them onto a smaller area of solar cells.

In this lecture photovoltaic concentration is covered dealing with both scientific and strategic aspects. It constitutes more the author's global views than a genuine revision work.

Accordingly, we present here five topics. We start with a description of the efficiency limits of ideal sun converters, and here, as homage to the Israeli development of solar thermal, we present the results of a yet unpublished theoretical comparison of the solar thermal and photovoltaic conversion processes.

Second, we describe the theoretical aspects of some solar optical devices used to achieve concentration.

Third, we describe the special solar cells needed for concentration.

Fourth, we describe the most important practical concentrators to date, with emphasis in our recently developed EUCLIDES prototype.

Fifth, we present several strategic considerations.

7.2.2   *Thermodynamic limits*

A solar energy converter is a device that receives (in a "receiver") solar radiation and produces useful power. In must supply some heat to the ambient and in addition, it emits some radiant energy. This is necessary because if there is a path for rays entering into the cell, by time reversal symmetry, the same path allows for the rays to escape. Possibly additional paths for escaping rays also exist, linking the receiver not only with the radiation source but also with the darker parts of the sky [1].

The nature of this escaping radiation defines the nature of the solar converter. If it is free radiation at a certain temperature, the temperature of the receiver, then the converter is solar thermal.

If the escaping radiation is a matter coupled (luminescent) radiation, characterized by a non-zero chemical potential, at ambient temperature, then the converter is photovoltaic.

In the ideal case this re-emitted radiation would be sent back to the sun, not allowing the escaping photons to reach darker parts of the sky. To achieve this, an ideal concentrator, or at least some optical device, must be used. That is why ideal conversion of solar energy is related with concentrators: the escaping radiation of a flat panel is isotropic and goes mostly to the darker sky.

A higher rate of useful work production by a solar converter occurs if the rate of irreversible production of entropy is zero, that is if the transformations are reversible, but neither process (solar thermal nor photovoltaic) is, except in certain cases, reversible. In reality both processes present a fundamental mechanism of irreversible entropy production. In the case of the solar thermal converters this irreversibility arises when the incoming radiation thermalizes to the receiver temperature. In the photovoltaic conversion, when it thermalizes to the luminescent radiation.

In addition, numerous additional irreversibilities take place in practical converters. In the solar thermal ideal case the heated receiver drives a Carnot engine, but in practice any real engine differs substantially from the Carnot ideal. In the photovoltaic case, the non-radiative recombination (often dominating the cell behavior), the limited mobility and the contact resistance are causes of irreversibility.

However, there is a drastic difference between a solar cell and a solar thermal converter. It is the fact that the cells are transparent to the photons with energy below the bandgap. This offers an advantage to the solar thermal converters, but one that can be compensated by using semiconductors of different bandgaps, or even better, by introducing intermediate levels within the bandgap [2]. Therefore, it is a better comparison to relate a solar cell with a solar thermal absorber having zero absorbance below a certain threshold and above another. This is presented in Fig. 7.1 that shows the output power normalized to the captured incoming power. Thus, what we are

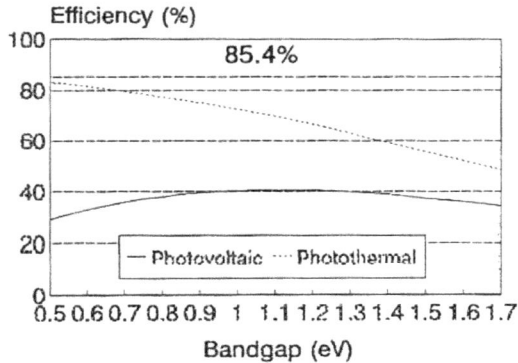

Figure 7.1:   Efficiency of an ideal solar cell and an ideal solar thermal converter with light absorption above E_g.

presenting is a technical efficiency. It is observed that the efficiency is lower in the photovoltaic case.

To reduce or avoid the irreversible entropy production associated with the thermalization of the incoming energy it is helpful to consider the use of an extremely narrow-band filter that allows the passage of monochromatic radiation. As such, this is the only radiation that can escape from the converter. In this case the solar thermal device and the photovoltaic device can produce exactly the same power output, which is presented in Fig. 7.2. In this case only the monochromatic radiation has been considered as power input.

In the two cases studied, a multiplicity of converters can be used, one for each wavelength. A configuration of it in the case of solar cells is presented in Fig. 7.3. In this case the efficiency is 86.8%. This efficiency is obtained by adjusting the voltage of each cell to maximum power point. This requires that every cell has its own independent circuit. In the case of solar thermal devices the temperature in the thermal receiver plays the same role as the voltage in the solar cell. Therefore, the 86.8% is obtained by adjusting independently the temperature of every absorber. However if all the absorbers were at the same temperature this would result in the case of a single absorber and a single Carnot engine. The efficiency if the

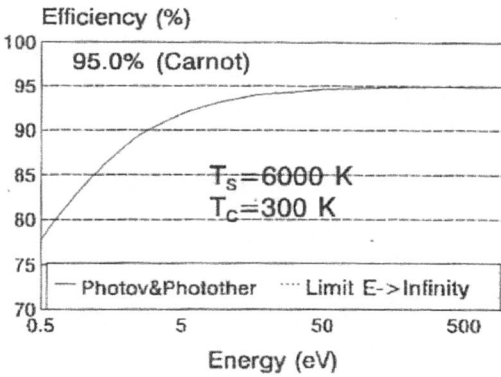

Figure 7.2: Efficiency per unity of filter bandwidth of an ideal monochromatic converter, either solar thermal or photovoltaic.

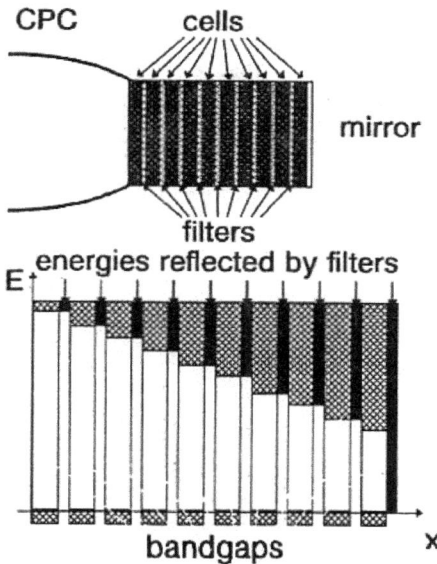

Figure 7.3: Stack of cells for maximum efficiency in photovoltaic conversion.

temperature is properly selected is 85.4% [3]. The loss of efficiency is not so high. In solar cells this simplification cannot be performed. However, a thermophotovoltaic device, that we shall describe below, can reach exactly the same result.

An interesting question is the following: Can a solar converter reach the Carnot efficiency between the gas of solar photons and the Earth temperature? Before answering, we need to discuss on what we consider to be the input power. To talk about Carnot efficiencies, the input power cannot be the sun power reaching the device, but the heat spent, and this is the radiation reaching the cell, less the radiation from the converter that reaches the sun again. We assume that the photons of energy not adequate to be absorbed by the converter are not absorbed by anything else and are sent back to the sun again. Under these conditions several devices can reach the Carnot efficiency, and therefore will not produce any entropy.

For ideal solar thermal converters, this will occur if the absorber temperature reaches the sun temperature. This only occurs if the work produced is negligible. This is coherent with the fact that a reversible operation is very slow.

For ideal solar cells, they must be covered by a monochromatic filter, and then they reach the Carnot efficiency if the cell is under zero current condition. Again the power is negligible.

Finally, we shall devote some words to the thermophotovoltaic converters [4]. A schematic diagram of one is shown in Fig. 7.4.

A thermophotovoltaic converter is a device in which the receiver is a thermal absorber that is heated by the sun. This heated absorber then emits thermal radiation that is converted into electricity by a solar cell. Since the radiation emitted back to the sun by this thermophotovoltaic converter is thermal radiation, we must classify it

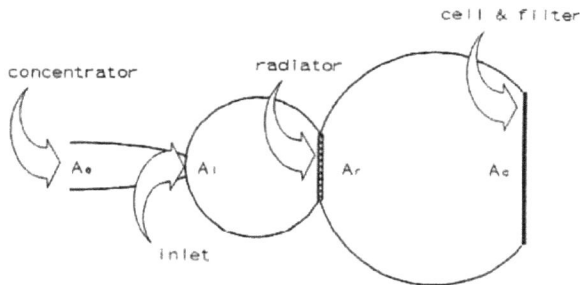

Figure 7.4:   Schematic of an ideal thermophotovoltaic device.

as a solar thermal device. The solar cell can be covered by a filter and operate near open circuit thus behaving as a Carnot engine. Under these conditions the efficiency of the thermophotovoltaic device is the same of that of the ideal solar thermal device, that is, 85.4% [5].

Thus, in summary, we can say that solar thermal devices and solar cells are theoretically equivalent if the necessary complexity is taken into account.

7.2.3   *Concentration optics*

In many cases, the optics used for photovoltaic concentration design is the kind simply derived from elementary geometric optics, which mainly uses the laws of refraction and reflection. Design, when not trivial is performed by ray-tracing codes based upon these laws. This is the case for the parabolic mirrors and, to an extent, for Fresnel lenses where the design tries to achieve, as far as possible, homogeneous illumination and insensitivity to misalignment. But even in such cases, a background knowledge of non-imaging optics is useful [6].

Non-imaging optics deals with bundles of rays and not with individual rays. It is a very useful tool for the design of systems that must transmit luminous energy. It allows for a very elegant formalization within the framework of Hamiltonian geometric optics.

A major class of results given by non-imaging optics is the determination of the maximum concentration that a given system can achieve. This is based on the conservation of the *étendue* — or multilineal invariant of the Hamiltonian optics — of a certain bundle of rays when it proceeds along an optical system.

Thus, may be derived the well-known formula stating that the limit of sun concentration achievable on Earth is

$$C = n^2 / \sin^2 \varphi_s \approx 46{,}747 \, n^2 \tag{7.1}$$

where n is the index of refraction of the medium surrounding the cell and φ_s is the sun angular semi-diameter. We present in Table 7.1 [7] the (approximate) absolute limit of concentration for several families of concentrators: assuming that theoretically they collect radiation

Table 7.1:   Theoretical and practical concentration limits.

Concentrator type	Max theoretical concentration source: 0.265°	Max practical concentration source: 1°
Circular reflective parabolic dish	*11687*	*821*
Parabolic dish with secondary (n = 1.49 for secondary)	*103783+*	*4800**
Square flat Fresnel lens square cell	*376*	*73*
Square flat Fresnel lens with secondary (n = 1.49)		1700*
Linear, flat Fresnel lens	*22*	*10*
Linear, arched Fresnel lens	*54*	*26*
Linear, parabolic reflector	*108*	*29*

+Totally ideal concentrator. *Optimal secondary for a Lambertian primary

only within the sun's semi-diameter, 0.265°, and; more practically, assuming that they are adapted to collect the rays in an angle of 1° in order to give some allowance to the construction and to the tracking.

A second class of results provides information about incompatible aims. For instance, in certain optical designs the achievement of homogeneous illumination is incompatible with the achievement of a large angular acceptance, while in others they are compatible.

A third class of results concerns the achievement of maximum values of the product of concentration-acceptance angle. For instance, certain focal distance leads to this maximum for every family of classical primaries (mirrors of any well designed shape, flat Fresnel lenses of any good design, curved Fresnel lenses in a large variety of shapes, etc.).

Besides these aids to design, non-imaging optics allows for real synthesis methods, one of them being the well-known string method discovered by Winston. But several others [6, 8–12], equally powerful, also exist, based on totally different principles. Unfortunately they are almost unknown by many non-imaging optics workers.

Most of the synthesis methods are two-dimensional. Real three-dimensional methods exist as a theoretical curiosity [11] but they cannot be used effectively so far. In consequence most designs are made for linear structures or for axial symmetry and in most cases the synthesis is done for meridian rays. Then the three-dimensional behavior of the concentrator is tested using ray-tracing methods and in some cases trial and error modifications are performed based on the intuition of scientists with good non-imaging optics backgrounds.

Among the concentrators based on such non-imaging optics design are many of the secondaries that are in use today. For instance, compound elliptical concentrators (CECs) [13, 14] and GVF-trumpets [11, 12] are very popular as secondaries. The "silo" secondary is less based on non-imaging optics synthesis and has the advantage of producing perfect homogeneity for angular acceptances not optimal but enhanced (with regard to the absence of the secondary).

A further area in which non-imaging optics may hold the key is in high concentration. For instance, thin concentrators for very high concentration have been recently invented and are being developed these days in order to produce a maximum angular acceptance-concentration product in the concentration range from $1000\times$ on, using an advanced extreme rays method (Miñano's method). Although it is possible to achieve the same result with classical primary-CEC combinations, the compactness of Miñano's approach provides, at (very) low cost, a range applications between laser and classical optical devices for optical communications and for illumination using LEDs.

Some of these new concentrators are illustrated in Fig. 7.5 showing the thickness to diameter ratio (T/D) and the fraction of energy (Tr) transmitted according to three-dimensional ray tracing.

For solar energy they can solve the problem of practical very high concentration ($1000\times$) devices for use with the expensive III–V single junction or tandem cells intended for very high efficiency. However, here material problems associated with high light fluxes have not yet been addressed owing to lack of budget, although they do seem to be soluble.

Another area where non-imaging optics has something to say is that of static concentrators.

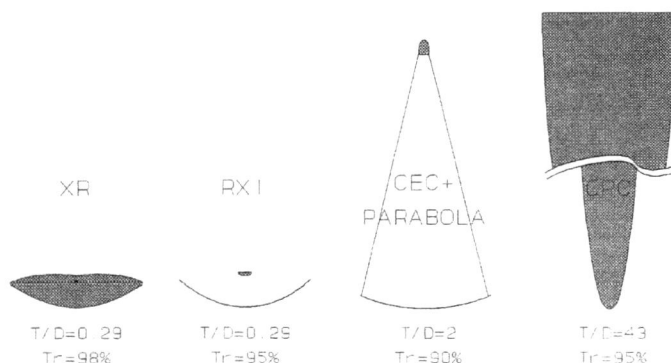

Figure 7.5:   Outlines of some concentrators designed with Miñano's method. Acceptance angle 1°. Index of refraction 1.5. Geometrical concentration 7387.

Static concentrators are concentrators with a very broad acceptance solid angle, so that they are able to collect rays from all the positions where the sun might be found during the year. Therefore they do not need to track the sun. Because of this broad acceptance they also collect diffuse light.

The concentration allowed by such systems is rather moderate [15]. The use of bifacial cells doubles it. Such cells were developed by us in the early 1980s [16] and were commercialized subsequently. (Several medium size plants were fabricated and a certain amount of commercialization took place in isolated homes).

The maximum concentration for static concentrators with bifacial cells is given by

$$C = 2n^2\pi/A \qquad (7.2)$$

where the angular acceptance area A is the area of the collected rays in the (p, q) plane of optical cosines. This concentration is only half if conventional single faced cells are used.

Clearly, immersing the cell in a transparent optically dense medium is a very good way of increasing the concentration. The reduction of the angular acceptance A is also a good option. A reduction of the angular acceptance will allow for an increased concentration but will reduce the collection of diffuse radiation, given by π/A. However, even if we want to avoid any reduction of the diffuse light capture it must be understood that large angle rays are usually reflected by the module surface and therefore they are not useful even if the optical system accepts them. In consequence it is most reasonable to design the optical system so that these rays are discarded and increased concentration can be achieved. Thus, taking into account that the index of refraction of typical materials is 1.5 the concentration achievable with a bifacial system that collects rays hemispherically $(A = \pi)$ is $2n^2 = 4.5$ but if we accept only the light within a cone of $70°$ $(A = \pi \sin^2 70°)$ then we can obtain a concentration of 5.1.

It is a common option to design the acceptance area so as to collect the direct sun during the whole year. The area in the (p, q) plane where the sun can be found at some time in the year is 1.549 (at sites of latitude $40°$). In consequence the maximum concentration of such a concentrator (with $n = 1.5$) can be of $9.12\times$. The fraction of diffuse light collected by such an ideal device is 49%. However to achieve this ideal situation is difficult. Most practical static concentrator designs collect slightly more diffuse light and achieve substantially less concentration.

The simplest, and so far most successful prototypes are modifications of Winston's compound parabolic concentrator (CPC), as that shown in Fig. 7.6, but other options have been explored. The main problem of the dielectric-empty CPC, besides its performance (which

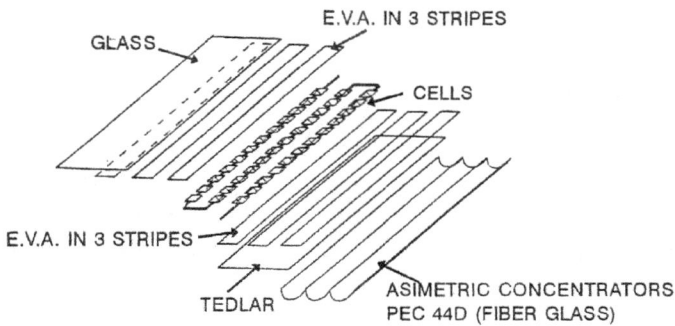

Figure 7.6: Linear static concentrating module using flat module encapsulation techniques.

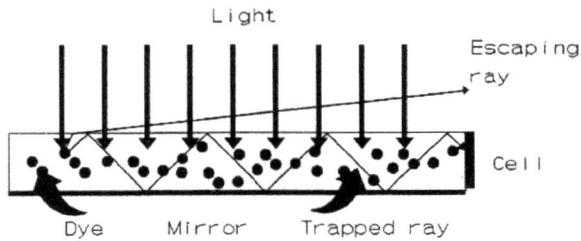

Figure 7.7:   Schematic of the operation of luminescent concentrators.

is not bad, but cannot be as good as one with included dielectric), is that it does not have a "high tech" look that can allow it to displace a flat module of equal performance.

Laboratory solutions have been demonstrated but, so far, without success in showing manufacturability. Again here it is mainly a matter of budget for the development.

It is worth mentioning that this requirement of a "high tech" look is achieved by luminescent concentrators [17]. These are transparent plastic sheets doped with a luminescent dye. They collect the light and re-emit it inside the plastic plate, where it remains captured and is transmitted to the edge of the sheet, by total internal reflection. There it is converted into electricity by solar cells. A schematic is shown in Fig. 7.7.

The main drawback of this approach is that the dye is not stable. Yet this device can be used for some consumer products such as clocks.

7.2.4   *Concentration cells*

Under increased irradiance (luminous power flux) any solar cell tends to produce a higher short-circuit current, closely proportional to this irradiance. The open-circuit voltage increases with the short-circuit current in a form approximately proportional to its logarithm, and so it does with the irradiance. The fill factor, that gives the maximum power output divided by the product of the short-circuit current and the open-circuit voltage, increases very slowly with the open-circuit voltage, in the absence of ohmic losses. The efficiency is then the product of the three parameters mentioned (short-circuit current, open-circuit voltage and fill factor). Therefore, in the absence of such losses the solar cell efficiency increases with the concentration level.

Unfortunately ohmic losses are not absent in any solar cell and they produce a drop in the fill factor that reduces the efficiency when the irradiance is very high. Any cell has a concentration at which the efficiency is highest; this concentration is that at which the ohmic voltage drop IR_s equals the thermal voltage kT/e (25 mV at ambient temperature) [18]. At concentrations above it the efficiency decreases. This explains why the series resistance of a concentrating solar cell must be much lower than that of a flat module cell.

To reduce the series resistance the solar cell must have a dense front contact grid. But very dense grids reduce the light entering the cell and produce an efficiency reduction, so that an optimization procedure is needed. In general, dense grids (10% of coverage) with narrow and thick fingers of good conducting metals are necessary for the higher concentrations.

In addition, concentration cells must be relatively small in size — not only for the purpose of reducing the series resistance but also because the handling of large currents produced by the cell under concentrated sunlight becomes difficult.

On the other hand, the concentration cell is only a part of the cost of the concentrating system, so that the effort in achieving high efficiency is justified. This higher efficiency may involve the undertaking of processing operations that are discarded in one-sun cells.

But under the heading of concentrating cells a large variety of options appear, depending on the level of concentration envisaged. Up to ten suns the screen printing technique used for one-sun cells, if properly optimized, is adequate. Above this level, the LGBG cell, developed by M Green in the UNSW [19] and industrialized by BP Solar, can operate up to some 30 suns. It is illustrated in Fig. 7.8.

Both cell types can take the shape of stripes, and can be used in conjunction with linear optics. A typical size may be some of $4 \times 10 \, \text{cm}^2$. In such cells it is the width (4 cm) that is the important dimension for the series resistance.

The LGBG cell has several additional advantages concerning efficiency. The most obvious is its metal fingers, which are buried in a

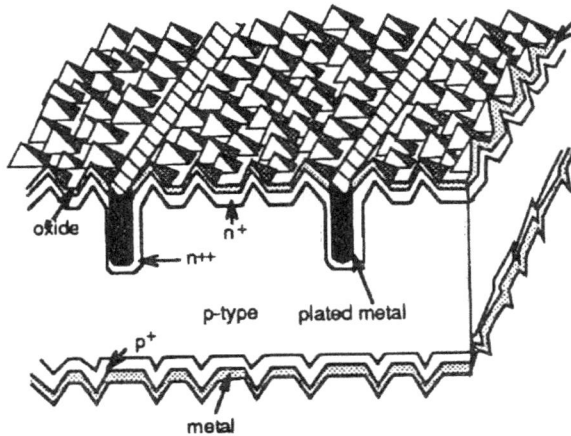

Figure 7.8:   Schematic of Laser Grooved Buried Grid (LGBG) cell.

laser groove and therefore they are both narrow and thick, resulting in low ohmic loss and low light obscuration. In addition the fingers are of electroless copper, silver covered, so that their conductivity is excellent.

In addition, there is a selective emitter, thick under the fingers and thin elsewhere. The thin emitter is the efficient part, but recombination in the metallic contact is reduced by having the emitter thick beneath it. Finally there is a p+ layer based on Al diffusion that reduces recombination at the back metallic contact.

Efficiencies in the vicinity of 18.5% are regularly achievable in fabrication [20].

For larger concentrations of 100–400× the cell must be much smaller, of $1\,\text{cm}^2$ or less, and squared or circular. In such cases the classical solution is to use a photolithographic grid, basically of silver (with other metals for contact and adhesion) with fingers $10\,\mu$m wide and 4–$5\,\mu$m thick.

For this range of concentration the most advanced cell is the PC cell, shown in Fig. 7.9 [21].

This cell is based on a high quality silicon structure that departs from the classical pn junction design, used for most cells. N and P dots are diffused in the cell rear, and are contacted together to interdigitated grids. In this way the cell recombination is highly

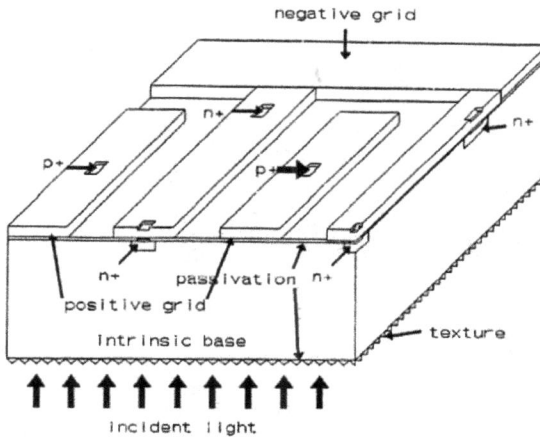

Figure 7.9: Schematic of the Back Point Contact (BP) cell.

reduced, and no grid is located on the front face. The bonding of this cell is not easy but has been solved using additional silicon wafers. Efficiencies in the vicinity of 25% at 200 suns are regularly achievable.

One of the important features of the PC cells, the absence of frontal fingers, has been matched to a couple of optical devices that reduce or remove the drawbacks created by the frontal shading.

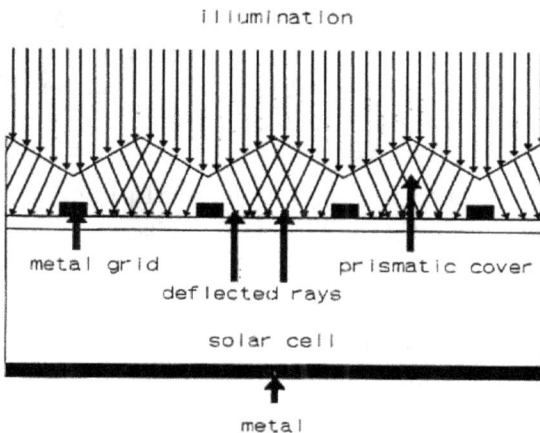

Figure 7.10: The Entech cover intended to avoid grid shading.

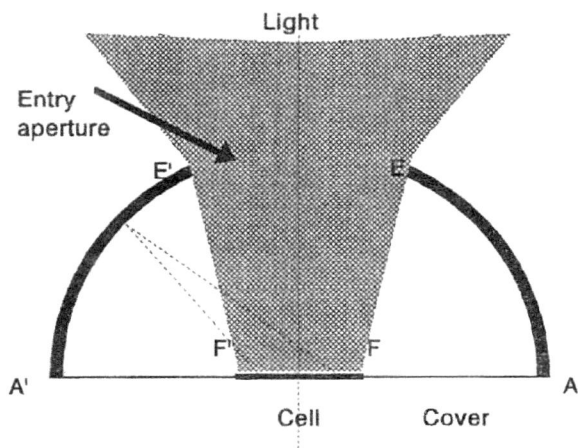

Figure 7.11:   Outline of the ellipsoidal cavity for light confinement.

One such device is the prismatic cover [22], developed by the American company ENTECH. A schematic of it is shown in Fig. 7.10. It is made from silicone rubber, capable of being shaped with a micro-mould. This cover deflects onto the cell that portion of the incoming rays that would otherwise be obscured by the fingers.

Another device, developed at our Institute, is the light confining cavity [23, 24]. This is a device that sends back to the cell most of the reflected rays. This improves the absorption not only of the rays reflected at the fingers, but also of those reflected by the cell surface. The best realization of this concept is the ellipsoidal cavity in which any ray from the cell is reflected back after a single reflection from an ellipsoidal mirror, as represented in Fig. 7.11.

Similar gains, of about 8% (relative) in the efficiency, have been observed with both types of optical device, In Fig. 7.12, we show an example of the gain with an ellipsoidal cavity.

So far we have been talking of silicon cells. III-V semiconductors, such as GaAs, are also of interest for concentration. They allow for higher efficiency with perhaps less technology complexity than the best silicon cells. In addition they allow for very high efficiency if tandem structures of several bandgaps are used.

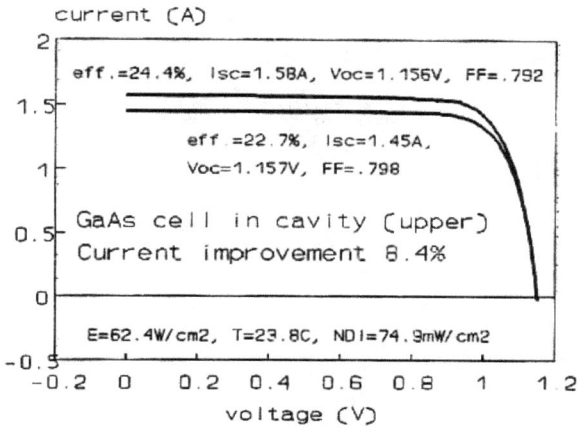

Figure 7.12: Current improvement in a GaAs cell located in an ellipsoidal cavity. E is the irradiance on the cell. NDI stands for the normal direct irradiance, almost that of the AMI.5D spectrum.

In its structure the GaAs cell is quite classic. There are two competing techniques for their fabrication: the metalorganic chemical vapor deposition (MOCVD) technique, which allows for very easy deposition of complex structures, and the liquid phase epitaxy (LPE) that produces better material but is less flexible.

Efficiencies in the vicinity of 28% have been produced by several laboratories at concentration of 200 suns. The problem is that this concentration is considered too low to be cost effective. By LPE the late professor Araújo, and his collaborators developed at our Institute GaAs cells with efficiency of 23.5% at 1300 suns [25]. Such cells are circular of area $0.1 \, cm^2$. We think they are appropriate for cost effective solutions when used with Miñano's high concentration devices [26].

Interest in using tandem cells is well known. These cells may be monolithic, so that the two semiconductors are deposited on the same semiconductor substrate. The complexity behind this endeavor is considerable. The two semiconductors must have almost the same crystal lattice constant and must have appropriate bandgaps, of 1.4

and 0.7 eV or so. In addition series connection between the two cells, with low series resistance is very challenging.

Better results have been achieved so far with a mechanical stack of a GaAs cell and a GaSb cell, which has given an efficiency of 32.8% at 100 suns [27]. But we think that this concentration is too low to be cost effective.

7.2.5 *Concentration systems*

The first modern photovoltaic concentrating panel, shown in Fig. 7.13, was developed by Sandia National Laboratories in the late 1970s [28]. The basic concentrator is formed by a point focus Fresnel lens that casts the radiation onto a circular cell of about 5 cm diameter. The concentration ratio is in the vicinity of 40×. A set of such elementary concentrators are attached to a beam that rotates about a horizontal axis (elevation tracking). This beam is mounted on a pedestal that rotates as a whole in azimuth. In later developments, the entire ensemble of cells is protected against excessive dust by a housing.

The tracking electronics of this system is based on a sensor that gives an error signal when the sun is out of aim.

Figure 7.13: Photovoltaic concentrator fabricated in the late 1970s in Sandia National Laboratories, Albuquerque, NM.

Cooling of the cells is achieved, again in later developments, by an aluminum extruded multifin heat sink. This Sandia Labs concept has been adopted by several manufacturers, including Martin Marietta, who installed about 350 kWp in Saudi Arabia [29]. Today some other companies are using it, as for instance Alpha Solarco, and Amonix [30] who have developed large panels in the 15–25 kW range.

While many other concepts were under design in those days, the next original concept that became a reality was the ENTECH concept [31]. Using it a plant of 300 kWp and several smaller installations have been installed.

This is a concentrator in which the cells are series connected in a linear row. The row is located under an arched Fresnel lens of linear focus, as shown in Fig. 7.14. Such arched lenses are strikingly insensitive to their precise position and allow a large angular acceptance. The concentration level is in the range of 10 suns, and in some cases screen printed cells are used.

Cooling is again via extruded heat sinks. A housing encloses the whole module.

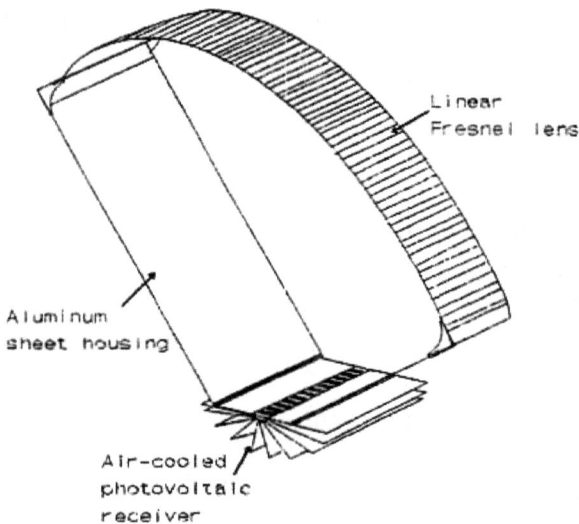

Figure 7.14: ENTECH concentrator. It has curved lenses with an axial focus, tracking the hour angle on a frame that tracks the sun's elevation angle.

Such modules are located on a frame that rotates following the hour angle. The frame as a whole rotates in elevation and rests on two poles at its ends. The tracking is similar to that of the Sandia prototype.

A concept, recently developed jointly by BP Solar and our own Institute (in an EU joint project in which the University of Reading and ZSW-Stuttgart also participated) is the EUCLIDES concentrator [32, 33]. Because we understand it in great detail we are going to explain it at somewhat greater length, so that arguments relevant to the design will be made manifest.

In the receiver module the concentration cells are interconnected and encapsulated as in flat modules. In the past concentration cells were delivered as components and it was the problem of the concentrator manufacturer to bond them etc. This required the use, as already mentioned, of a rather expensive "housing" to protect the cells. Now, a completely encapsulated concentrating module, professionally bonded at the cell factory, can be easily handled by the concentrator manufacturer.

However this module is a more sophisticated product than a flat module because of several reasons. Specifically, the encapsulant must be able to withstand high irradiance and high temperatures (which could occur in the case of an accident); moreover, the module must have an extremely low cell interconnection series resistance, and it must permit excellent thermal contact with the heat sink, while preserving very good electrical insulation with the surrounding — even at high voltage (2000 V dc). In addition the modules are fitted with a bypass diode for protection against hot spots.

In EUCLIDES the concentrating optics is mirror-based instead of the Fresnel lenses used previously in all (successful) photovoltaic concentration developments. When comparing lenses vs. mirrors there are pros and cons for each option, but two reasons persuaded us to use mirrors. One is the fact that mirrors, and not lenses, allow the use of one axis tracking systems, which are less costly than two-axis trackers. The other reason is that lenses must be protected from dust on their rear face, so that an expensive "housing" is needed.

In addition the relative success of the concentrating solar thermal options in the USA proves that the mirror option is promising as a cost effective option.

Once this option was decided the profile was optimized using non-imaging optics, so as to allow for the largest range of manufacturing errors for a given level of concentration (geometrical concentration of 32×). Since this profile was not available on the market, and because of the precise profile that was required (which we are not sure can be achieved with glass), a new fabrication technology was developed. It used a very well protected 3M silvered layer, laminated onto a thin sheet of aluminum to which the shape was subsequently given with high precision and then affixed to a supporting aluminum frame. Presently the mirrors are fabricated by JUPASA, under the technology of the IES/UPM — one of the EUCLIDES consortium. A sketch is shown in Fig. 7.15, showing the narrow strip of light, with high translational symmetry.

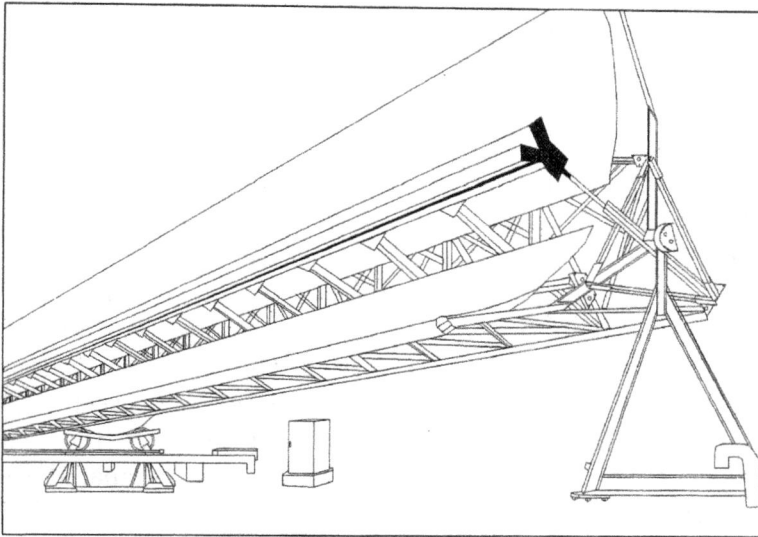

Figure 7.15: View of the EUCLIDES concentrator with the cells illuminated by concentrated sunlight.

The mirrors have an efficiency of 89% and an outstanding quality so that most of the light is cast on a strip of about 2 cm width within the 4 cm that the cell has for receiving it.

This good optical behavior is quite essential for the success of this technology because in this system all the cells are connected in series so that a translational inhomogeneity of the illumination would be reflected in loss of fill factor. In this way the system can become twisted by the effect of the wind, etc., and some moderate mounting misalignment can be allowed without affecting the electric output.

The explanation for the mirror's good behavior is that the flexible sheet technology used allows for an excellent reproduction of the theoretical profile, something that is doubtful with glass mirrors; and that the theoretical design follows the rules of non-imaging optics for minimizing the width of the strip of light.

One of the common questions relating to solar cells under concentration refers to the presumed difficulty of keeping them cool. It seems natural that the cell under concentration will become hotter, the more so if we remember that with similar optics a fluid is brought to some hundreds of degrees centigrade. This is often thought to be an almost insurmountable obstacle for concentration, for it is well known that solar cells have reduced efficiency when their temperature is increased.

However, a properly designed heat removing system can keep solar cells at reasonably low temperature. Moreover, for the level of concentration and the size of the receiver (i.e., the cells) we are considering here, cooling can be achieved by passive heat transfer to the ambient, making use of the natural air convection in a finned heat exchanger. To this end the EUCLIDES concentrator uses a finned aluminum cooler that has proven to be able to keep the cells in the range of 30–40°C above ambient, depending on the climatic conditions (wind and irradiance). This may be compared to the case of the cells in a flat module (that typically are 15–25°C above ambient).

But, of course, this cannot be achieved without a good technology. The removal of heat is achieved by transport to the ambient air, which is subsequently heated and set in motion by buoyancy forces. This requires both an effective transference of heat to the air — which is achieved by increasing the surface area with a vertically oriented multifin structure — and an effective transference of the heat from the cell to the fins by heat conduction. The latter is achieved using aluminum as the forming material of the cooler and optimizing the shape of the different elements: critical features being the fin thickness and spacing. A physical modelling of the problem and an optimization was developed at the IES/UPM, in order to minimize the temperature drop for a given weight of aluminum. As a result, the extrusion technology, previously used by us, and by our competitors, was discarded and substituted by another low cost technology that uses about half the amount of material of the earlier technology. This cooler produces a temperature drop of 30°C in the absence of wind, for the maximum power expected on the cooler.

Another aspect to consider is the reduction of the thermal drop in the module-heat sink interface. This is reduced to 1°C with the help of a thermal paste that we developed. Another temperature voltage drop, of about 6°C, is produced in the cell-module bottom interface due to the electrical insulation that is also a thermal insulation.

In this way the maximum temperature drop is in the range of 37°C, although in most cases the actual drop is substantially less because of the lower solar power input and the effect of the wind.

The tracking structure is one of the most important elements of a system for concentration because it constitutes the main penalty to be paid by the reduction of solar cell area, in terms of cost. Many options are possible here, but developing each one of them, for cost effectiveness, is not a trivial matter. The solution adopted for EUCLIDES is one that we consider adequate, and no better solution is apparent to us today.

The option we have adopted, was inspired by the well-developed solar thermal solution of California (Luz Solar technology). It is a horizontal one-axis tracking system using mirror optics. It is thought that one-axis solutions are cheaper than two-axes tracking mechanisms.

A very important argument to further support this choice is the fact that in the horizontal one axis-tracking configuration the cells are easily interconnected (in series) so that the photovoltaic element has a high voltage output. This is very important because it avoids costly field cabling and interconnection, and allows for a less costly inverter directly connected to the array. Thus, the array output is already ac.

Once this solution is adopted we must insure a translational homogeneity of the strip of concentrated sunlight (see Fig. 7.15). For this purpose the mirrors must be closely spaced forming an almost continuous mirror surface along the entire length of the surface. This means that few intermediate supports should be used.

With all this taken into account the solution adopted, represented in Fig. 7.15, consists of a lightweight reticulate long supporting beam (just like an electric tower placed horizontally) that rests in a central tracking wheel and two end poles, The modules, attached to the heat sinks, are mounted on appropriate arms issuing from the beam, which similarly supports the mirrors.

In both the module and mirror mounting enough degrees of freedom are left for easy adjustment of the focus. Special attention has also been paid to easy and fast assembly.

The entire beam is rotated by the wheel that is turned with a cable by a carriage moved by a worm screw. The sun is accurately aimed at by open-loop electronics, also fitted with some learning strategy. In this way the system can be installed with an arbitrary azimuth (NS gives the most energy output) that needs not to be very precisely determined, thus reducing installation costs.

The entire supporting and tracking structure was designed using finite-element codes, where all the stresses, including wind stresses were simulated. Ray tracings were then performed on the deformed structure to ascertain that the rays will fall into the mirrors even after the deformations expected for wind speeds up to 30 km/h — although the structure has elements to withstand much stronger winds but without maintaining operational sun-tracking in such circumstances.

Concerning the structure that we consider appropriate, it will be longer than that installed in the prototype (of 72 or 84 m long instead of the 24 m prototype). This is basic for several reasons, resulting in costs. The prototype system so far installed is of a size smaller than the projected one because of land and funding availability.

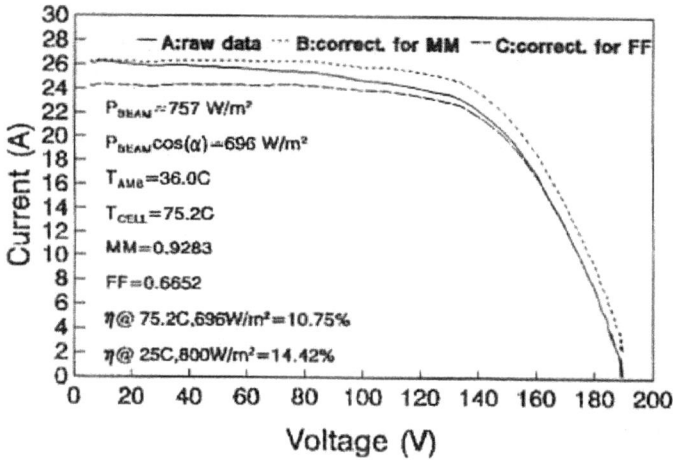

Figure 7.16: IV curves of the 28 modules (336 cells) in series in the EUCLEDES prototype array.

In consequence its steel weight is too high, because no smaller standard steel beams existed in the market. This is one of the main reasons for the larger size that is projected for the tracking structure. The mechanical behavior of the real size concentrator will be tested in the present prototype by applying artificially enhanced stresses.

The IV curves of the array as it is now, with 28 modules totaling 336 cells in series, are shown in Fig. 7.16. A certain mismatch is apparent but it has been determined to be entirely caused by the current dispersion of the cells at one sun. (No sorting was performed at the factory due to the small number of cells manufactured.) No light inhomogeneity mismatch has been detected so far.

A demonstration plant of $3479 \, m^2$ aperture (sixteen 72-m long arrays, with a concentration of $38 \times$) will be installed with EC support on the island of Tenerife, in the Canary Islands. A provisional cost analysis for this plant is given in Table 7.2, in comparison with the actual costs of a flat panel plant.

The cost of \$3.88/Wp indicated in the table is in the range (\$3.40–5.00/Wp) of those costs considered as competitive for independent power producers in the PVUSA program, and is even approaching the

Table 7.2: Cost per watt peak installed of a flat array plant, and a concentrating plant based on the present EUCLIDES technology after the fabrication of 10 MWp.

Cost in US $	TOLEDO 1 MWp flat module	EUCLIDES future	
	$/Wp	$/Wp	$/m^2
Modules	5.40	0.46	53
Mirrors	0.00	0.42	48
Heat sink	0.00	0.61	70
Module mounting	0.42	0.11	13
Installed module	5.82	1.60	185
Site preparation	0.22	0.21	24
Struct. mount & transp.	0.44	1.59	183
Struct. civil & transport.	0.66	1.80	208
Wiring & dc distribut.	0.44	0.06	7
Inverter, transfo., etc.	0.42	0.42	48
Power conditioning	0.86	0.48	55
TOTAL	7.34	3.88	447

Note: For concentrator the costs are also given per square meter of aperture. A substantial cost reduction seems apparent. Further cost reductions are expected in particular by increasing the mirror aperture.

range ($2.70–3.80/Wp) for production of PV electricity by utilities in high value applications [34].

7.2.6 *Strategic aspects*

The potential of solar cells to give higher efficiency under concentration is often mentioned as a main motivation for the use of this technology. This increase of concentration is due to the physical aspects already studied and to the fact that more care is often put into manufacturing concentration cells. This is permitted by the fact that as the concentration solar cells represent only a fraction of the concentrating system cost, the concentration solar cell may be expensive and therefore more carefully made, and this is a second reason for higher efficiency.

We think that the efficiency argument is not the strongest argument. In reality this efficiency advantage is a must to compensate the losses inherent to the optical system and to efficiency loss due to a higher temperature operation. Nevertheless, despite these losses, concentrating systems have often turned out to be more efficient than flat module systems [34].

Thus, if we are asked to summarize in a few words what is the motivation of concentrating photovoltaics, it must be answered that it is not efficiency but the reduction of cost by reducing the area of expensive solar cells.

It is worthwhile to mention here that a severe shortage of hyper-pure silicon has become visible on the horizon. This is due to the fact that today's photovoltaic industry is using the rejects from the microelectronics sector, to the extent of some 10% of the latter's consumption. However, in the microelectronics industry there is a tendency to reduce rejects. Thus, with high probability the cost reductions expected in the silicon photovoltaic manufacturing sector will be halted due to the increase in price of hyper-pure silicon. So we might even experience a photovoltaic module price increase.

A matter that is raised every time that photovoltaic concentration is examined is the loss of diffuse light. This is certainly the case for high concentration devices, which collect only the rays in the vicinity of the solar disk and focus them onto a small cell area. Such concentrators require sun tracking in order to keep the cell in focus, although this is not the case for the low concentration static concentrators.

Obviously, those concentrators not collecting the diffuse light experience a reduction in the electric output due to this loss. This is equivalent to a loss of efficiency. However, this loss is compensated, sometimes with advantage, by the need to use tracking. Tracking keeps the sunlight collecting area higher than in the case of static flat modules, which in the morning and afternoon hours receive the rays cosine and reflection losses. Obviously the use of tracking with flat modules would give them a comparative advantage, thus enhancing the apparent efficiency of the latter. If such tracking were to be used it is doubtful that concentrating approaches could exceed the operating efficiency of flat module fields. But the advantage for concentrators, that is strong, remains in the lesser use of expensive cells.

The fact that tracking concentrators only collect the direct beam suggests that their use is limited to clear climates. This may be wrong again. The yearly direct radiation in a two axis tracking plane is not so different from the global radiation on a latitude-tilted stationary plane for a very wide variety of climates (although not for some, such as wet equatorial climates). What is more of a fact is to state that in certain locations the availability of solar radiation, both direct and diffuse, is smaller than in others.

But, of course, if we are going to install a solar generator with the aim of being cost competitive with fuel, then we must go to locations with high radiation and in the case of tracking concentrators, with high direct radiation. If the installation has other aims (and this is most often the case today) then the cost argument loses value. To some extent this explains the small deployment of concentrating approaches today.

With respect to static concentrators, as they behave very much as flat modules, they share the advantages of them in many respects. The great disadvantage is that the reduction of cell area is moderate so that the cost advantage is probably small.

Concentrators have experienced a very modest commercialization. Beyond few demonstrations performed in the early eighties in the USA and in Saudi Arabia (using US technology), and the present Spanish demonstration, no real commercialization has occurred.

There are several reasons that explain it. On the one hand, except for a few demonstrations mainly intended to drive the flat module industry, the current market for photovoltaics is limited to small generators in scattered applications where the electricity is needed at almost any cost (professional or rural 3rd world applications). A different market, that of grid connected houses, is linked to ideological positions and there the costs and even performance means less. Architecturally aesthetic integration is much more important.

While these markets might become important for static concentrators, which, however, are not yet technically mature, they are probably not appropriate for tracking concentrators.

Generators based on tracking concentrators are, and will probably be, larger and not suited for many of the small scattered applications of today. Nor can they be easily integrated into buildings with the same beauty that can flat modules. They must compete more in the cost arena.

But beside this difficulty from the demand side, there is also one from the other side: the lack of concentration cells. Concentration cells are mature at the laboratory level, but due to the lack of market, not at the industrial level. Therefore they are very expensive and destroy most of the advantage implied in the reduction of cell area.

Consider that a flat module cell costs today in the vicinity of $0.04/cm^2$, while the price for a typical concentration cell (very variable depending on complexity) is in the range $0.5–1/cm^2$. A concentration of at least 12 is therefore needed merely to balance these costs. In reality the cost of such concentration cells is not a manufacturing cost, it is just an indirect cost of keeping in operation an R&D structure.

To keep operating a production line of concentrating cells, equivalent to say a 500 kWp line of flat modules, we must have a market of roughly this size times the concentration factor. If this factor is 10 then the market must be 5 MWp, if it is 100, then the required market is 50 MWp. Such markets do not exist today.

Another option is to use a line of fabrication of flat module cells to produce a concentration cell that is only a modification of the main product. With the technology of screen printing normally used this is restricted to a concentration of about 10. With the LGBG technology used by BP Solar, 30 or more can be easily reached. Most probably this is the way of starting.

Finally it has to be understood that no real marketing effort has been done so far for concentrators. The companies promoting this technology have been too close to the development side and have not had the skills (nor the funds) for marketing action. There is certainly a niche for a cheap concentration product, probably higher than for the flat module, but different, and this niche remains virgin. According to "Photovoltaic Insiders Report" [35], This situation might be drastically changed due to the involvement of the oil giant, British Petroleum (through BP Solar), in the concentration field with the commercialization of the EUCLIDES technology. For this involvement, an agreement has been signed between our UPM (who developed the optical and mechanical elements) and the company for world commercialization.

What are the characteristics of the niche market mentioned above?

If we compare it with the flat modules of today and most probably of the next 10 years, we can expect lower costs but a larger minimum size. Thus, concentration technology might be appropriate for 3rd world village electrification, for irrigation, and in some specially good climates for supporting electricity generation in certain cases.

If we compare with the solar thermal option what we can say is that we do not see why the concentrating photovoltaic option cannot be of the same or lower price. Efficiencies are not so different but simplicity of operation is better. Yet the modularity is much smaller so that its marketing should be much easier. This important advantage in modularity comes from the fact that efficient solar thermal converters are rather large — in the range of 30–100 MW — compared to less than 0.05 MWp for concentration photovoltaic generators.

7.2.7 *Conclusion*

In conclusion, we want to stress that to our understanding, photovoltaic concentration is ready for an important deployment in the near future. It is true that the utilization of this technology departs from the usual dreams of photovoltaic promoters, for perfect integration and full decentralization. But we think that prices neighboring those of conventional electricity in the sunbelt will be within the reach of this technology.

7.2.8 *References*

[1] A. Luque and A. Martí, *Phys. Rev. B.* 55 (1997) 6994.
[2] A. Luque and A. Martí, "Increasing the efficiency of ideal solar cells by photon induced transitions at intermediate levels", *Phys. Rev. Lett.* 78 (1997) 5014–5017.
[3] A. de Vos, *Endoreversible Thermodynamics of Solar Energy Conversion*, Oxford Univ. Press, Oxford, 1992, p. 81.

[4] R. M. Swanson, *Proc. IEEE* 67 (1979) 694–698.

[5] A. Luque, in M. Prince (ed.), *Advances in Solar Energy: An Annual Review of Research and Eevelopment,* vol. 8, Chapter 14, American Solar Energy Society, Boulder, 1993.

[6] W. T. Welford and R. Winston, *The Optics of Non-imaging Concentrators,* Academic, New York, 1978.

[7] A. Luque, *Solar Cells and Optics for Photovoltaic Concentration,* Adam Hilger, Bristol, 1989.

[8] J. C. Miñano and J. C. González, *Appl. Opt.* 31 (1992) 3051–3060.

[9] R. Winston and W. T. Welford, *J. Opt. Soc. Am.* 69 (1979), 532–536.

[10] R. Winston and W. T. Welford, *J. Opt. Soc. Am.* 69 (1979) 536–539.

[11] J. C. Miñano, *J. Opt. Soc. Am. A* 2 (1985) 1826–1831.

[12] J. C. Miñano, *Appl. Opt.* 24 (1985) 3872–3876.

[13] X. Ning, R. Winston, and J. O'Gallagher, *Appl. Opt.* 26 (1987) 300–305.

[14] J. C. Miñano, J. M. Ruiz, and A. Luque, *Appl. Opt.* 22 (1983) 3960–3965.

[15] A. Luque, *Solar Cells* 3 (1981) 355–368.

[16] A. Cuevas, A. Luque, J. Eguren, and J. del Alamo, *Solar Cells* 3 (1981) 337–340.

[17] A. Goetzberger and V. Wittwer, *Solar Cells* 4 (1981) 3–23.

[18] E. Sánchez and G. L. Araújo, *Solar Cells* 12 (1984) 263–267.

[19] M. A. Green, A. W. Blackers, S. R. Wenham, S. Narayanan, M. R. Wilson, M. Taouk, and T. Spitzalak, *Proceedings of the 18th PV. Special Conference,* IEEE, New York, 1985, pp. 39–42.

[20] T. M. Bruton, K. C. Heasman, and J. P. Nagle, *Proc. 12th EC PVSEC,* Stephens, Belford, 1994, pp. 531–532.

[21] R. A. Sinton, Y. Kwark, P. Gruenbaum, and R. M. Swanson, *Proceedings of the 18th Photovoltaic Specialists Conference,* IEEE, New York, 1985, pp. 61–65.

[22] M. J. O'Neill, US Letters Patent #4711972, 1987.

[23] J. C. Miñano, A. Luque, and I. Tobías, *Appl. Opt.* 31 (1992) 3114–3122.

[24] A. Luque, J. C. Miñano, P. Davies, M. J. Terrón, I. Tobías, J. Alonso, and J. Oliván, *Proceedings of the 22th IEEE Photovoltaic Specialist Conference,* IEEE, New York, 1991, pp. 99–104.

[25] J. C. Maroto, A. Martí, C. Algora, and G. L. Araújo, *Proceedings of the 13 European Photovoltaic Solar Energy Conference,* Stephens, Belford, 1995.

[26] J. C. Miñano, J. C. González, and P. Benítez, *Appl. Opt.* 34 (1995) 7850–7856.

[27] L. M. Fraas, J. E. Avery, V. S. Sundaram, V. T. Dinh, T. M. Davenport, J. W. Jerkes, J. M. Gee, and K. A. Emery, *Proceedings of the 21st PV. Special Conference,* IEEE, New York, 1990, pp. 190–195.

[28] E. L. Burgess, D. A. Pritchard, and R. D. Nashby, *Proceedings of the 13th PV Special Conference,* IEEE, New York, 1978, pp. 1121–1124.

[29] A. Salim and N. Eugenio, *Solar Cells* 29 (1990) 1–24.

[30] V. Garboushian, D. Roubideaux, and S. Yoon, *Proceedings of the 25th Photovoltaic Specialists Conference,* IEEE, New York, 1996, pp. 1373–1376.

[31] M. O'Neill, A. McDanal, R. Walters, and J. Perry, *Proceedings of the 22th Photovoltaic Specialists Conference*, IEEE, New York, 1991, pp. 523–528.
[32] G. Sala, J.C. Arboiro, A. Luque, J. C. Zamorano, J. C. Miñano, and C. Dramsch, *Proceedings of the 25th Photovoltaic Specialists Conference*, IEEE, New York, 1996, pp. 1207–1210.
[33] A. Luque, G. Sala, J. C. Arboiro, T. Bruton, D. Cunningham, and N. Mason, *Prog. Photovolt.* 5 (1997) 195–212.
[34] G. Jennings, B. Farmer, T. Townsend, P Hutchinson, T. Reyes, C. Whitaker, J. Gough, D. Shipman, W. Stolle, H. Wenger, and T. Hoff, IEEE, New York, 1996, pp. 1513–1516.
[35] Photovoltaic Insiders Report, XVI(1) January 1997, pp. 1 and 3.

7.3   Discussion following Antonio Luque's presentation

Roy: The concept of concentrator PV and small demonstration projects have been with us for almost the same amount of time as non-concentrator PV. And yet the latter enjoys an annual market of some \$1 billion whereas concentrator systems are almost non-existent. What, in your opinion, has held back the development of concentrator PV?

Luque: I believe there are two main reasons — and they are related to each other: First, the minimum size of a concentrator system is too large to enable it to enjoy anything but a specialized market. By comparison, the standard 50 W non-concentrator module has enormous marketability. Secondly, concentrator PV cells were too expensive because special technology was needed to develop them and the market was not there.

I believe that in Madrid we have overcome this second problem thanks to our having been able to develop — together with BP Solar — a concentrator cell based on BP's standard non-concentrator technology (which was, itself, based on one of Martin Green's patents). BP already enjoys a large part of the PV market and the prospect of their being able to manufacture concentrator cells without the need to re-tool their factory will, hopefully, provide concentrator technology with commercial prospects that had hitherto been impossible.

Now once you have a large company with the ability to mass-produce concentrator cells the relatively large size of a basic module becomes less of an obstacle.

This situation may be contrasted with the manner in which the US DOE attempted to develop concentrator technology in the 1970s. They encouraged a number of companies to produce small (i.e., of a few kW) demonstration systems and then drew the premature conclusion that concentrator PV systems are unreliable. The point is that you cannot expect reliability from one-off systems. Reliability comes from the learning experience associated with correcting many replicas. I hope that with our EUCLIDES project we shall be able to demonstrate both reliability and cost-effectiveness.

Licht: Can you remind us of the size of the EUCLIDES unit?

Luque: The basic design unit is 25 kW. It is 72 m long and 3.06 m wide. [But the one that will actually be built will have a length of 84 m and a nominal power rating of 28.5 kWp [post-symposium communication by the speaker].

Mancini: In the US, non-concentrator PV is moving more in the direction of the demand side of the electricity meter. Is this fact of relevance for concentrator PV?

Luque: The main reason for this development is the high cost of PV compared to conventional methods of power generation. All the time that these costs remain high — and I think that for non-concentrator PV they will remain so for the foreseeable future — then demand side applications (of which there are many) will remain the sole market. On the other hand, I believe that with concentrator PV we have the possibility of competing with conventional power generation — particularly in situations for which conventional technology is unusually costly (remote regions, etc.). Now that is not to say that I believe that one technology will win out in the long run. On the contrary, non-concentrator PV will find a market for large public building facades and other architectural purposes whereas concentrator PV will be the preferred solution for serious power production.

Faiman: Why did you choose parabolic trough rather than dish geometry for your reflector?

Luque: Because BP's cells are more suited to linear arrays. If we were to pack them into square areas there would be considerable wasted area where the contacts are located and the achievement of uniform illumination would be problematic.

Question: In the cost break-down table you showed for the EUCLIDES project why does your inverter cost only half as much as a conventional inverter?

Luque: This is a matter of terminology. The inverter itself is of comparable cost to most other PV inverters but I have included the field cabling costs which are essentially zero for the EUCLIDES system.

Question: Perhaps one reason that concentrator systems are not more widely used has to do with the relative lack of direct beam insolation in many places?

Luque: Actually I think this lack is more perceived than real. For most places whose solar radiation statistics I have examined — including places in Germany — the annual direct beam totals are usually larger than the corresponding global figures. Of course a tracking surface would receive more global radiation than direct beam — typically 30% more. I am, therefore, surprised that so few tracking non-concentrator PV systems have been installed. I feel sure that the additional energy would more than compensate the additional cost of tracking.

Appelbaum: 20 years ago we were not able to predict the way the PV market would develop. We thought that module costs would decrease at a much faster rate than they did (50¢/Wp by the year 1990!) and we did not foresee that the bulk market would be for small, individual modules. I therefore think that, today too, we should not expect to be able to foresee the direction that concentrator PV will take in the coming 20 years.

Luque: I tend to agree with you, particularly since there are many technical problems (e.g., heat removal, cell stability, etc.) that must be overcome. However, when they are overcome we shall see a quantum jump from 50 W modules to 25 kW modules — in the case of EUCLIDES. This will open up completely new classes of markets. But even if EUCLIDES does not succeed in doing this, another project will because BP have proved that low-cost concentrator cells can be mass-produced. The availability of such cells together with all the other giant market forces at work convince me that concentrator PV has a rosy future.

7.4 Integrated high-concentration PV: Near-term alternative for low-cost large-scale solar electric power (Eng. Vahan Garboushian)

A Keynote lecture prepared by Vahan Garboushian (Amonix Inc., 3425 Fujita Street, Torrance, CA 90505, USA)

*Editor's note: This paper was originally intended as a keynote presentation for the 7th Sede Boqer Symposium on Solar Electricity Production. Unfortunately, at the last moment Mr. Garboushian was unable to attend. A polished version of the paper was subsequently published in *Solar Energy Materials and Solar Cells*, Vol 47, Vahan Garboushian, Dave Roubideaux, Sewang Yoon, Integrated High-Concentration PV Near-Term Alternative for Low-Cost Large-Scale Solar Electric Power, pp. 315–323, Copyright Elsevier (1997). The following is the original text except for the figures, which have been reproduced, with appreciation, from the Elsevier version because of their superior quality.

7.4.1 *Abstract*

Large-scale photovoltaic electric power generation deployment and utilization is no longer dictated by limitations in technology, but rather by the economics of PV systems vs. other renewable or traditional options. This paper describes a near-term alternative option for cost-effective solar electric power generation based on a novel sunlight concentrating technology: integrated high-concentration PV (IHCPV). The advantages of high-concentration systems have been well analyzed, but development was constrained by the lack of solar cells capable of withstanding the rigors of concentrated sunlight. The development of a stable, high-concentration back-junction, point-contact cell, by Amonix, paved the way for high-concentration system development. System designers had to insure that the cost savings inherent in concentration systems through the reduction of costly solar cell content were not over-shadowed by the ancillary costs of structure and tracking elements used in concentrating arrays. The IHCPV system has met these goals. Economic factors specific to the IHCPV system are presented including (1) low cost of entry, (2) enhanced energy production, (3) reduced land utilization, and (4) accelerated benefits of volume production.

7.4.2   Background

PV technologies have proven their maturity for large-scale applications. However, market penetration for photovoltaic systems has been largely constrained by price. All PV manufacturers seek to reduce manufacturing expenses in order to bring the cost per watt of their systems below the threshold levels needed for acceptance by different markets, because each \$/W reduction in PV prices opens up new market segments with great potential.

7.4.3   Cost reduction strategies

The principal strategy for reducing cost is to reduce the material and/or manufacturing expenditure for those system elements which contribute the most to total system pricing. The principal element in any photovoltaic system is the high-priced solar cell; generally, 50% or more in typical systems. The solar cell is the economic key to any photovoltaic system.

The predominant material used for today's solar cells is crystalline silicon. It constitutes a mature technology where labor costs have been minimized in comparison to material costs. Here it is doubtful that any major technical breakthroughs can be achieved which will dramatically increase its performance or reduce its manufacturing expense. Performance for crystalline silicon solar cells is approaching its practical and technical limit. Nevertheless, despite its good performance and well-understood manufacturing processes, traditional crystalline silicon solar cell technology remains relegated to uses which can support its high price.

Reducing the cost of high-priced silicon PV material is the primary focus of the majority of differing PV technologies. All approaches strive to reduce the silicon material content or move to lower-cost alternative materials.

Most of the PV industry is pursuing the strategy of reducing the thickness of the cells themselves by using very thin films of silicon or alternate materials. Most of these approaches are relying on technical breakthroughs in performance coupled with extensive capitalization. The expected results are still several years away.

Concentrator systems, which concentrate sunlight use a highly cost-effective approach of reducing the area of expensive cell material required to generate a given amount of electricity. This technology, described below, is available today.

Typical, one-sun flat-plate solar modules have their entire sun-receiving surface covered with costly silicon solar cells and usually operate at a fixed tilt to the Earth. Concentrating system designs fall into two major categories: low-concentration ($10\times$–$50\times$) and high-concentration (i.e., $>100\times$). Maximum performance and associated cost benefits can only be achieved at high concentration. High-concentration Fresnel lens concentrating systems offer significant cost savings by using inexpensive flat, plastic Fresnel lenses as an intermediary between the Sun and the cell (see Fig. 7.17). These magnifying lenses concentrate sunlight 250–500 times on a relatively small cell area. The concentrating system's unique mode of operation reduces the silicon cell area required by an amount approximating its concentration ratio.

Although the high-concentration photovoltaic concept had been theoretically studied for many years, it was not considered commercially practical because of the lack of solar cells capable of withstanding the punishing environment generated by highly concentrated sunlight.

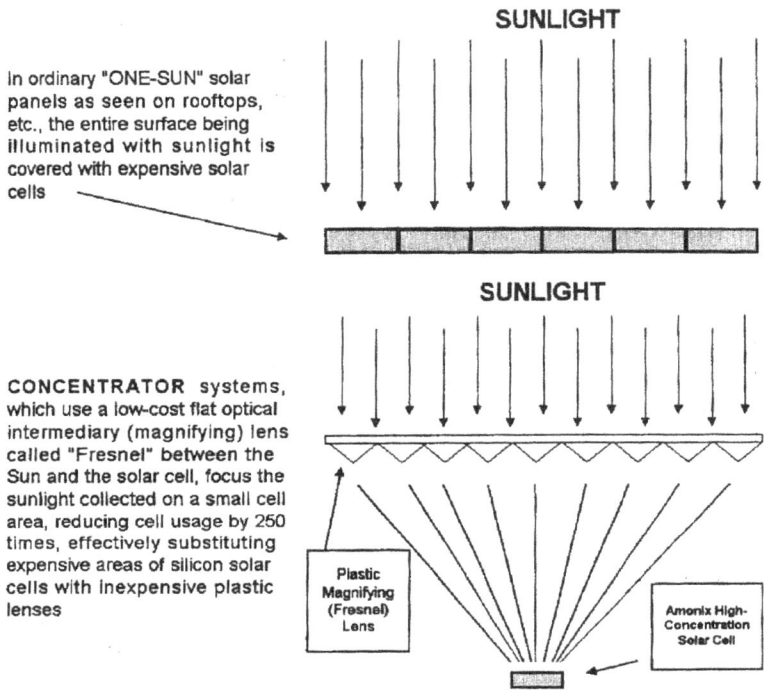

Figure 7.17: Graphical representation of operation of flat-plate vs. concentration solar systems.

7.4.4 *High-concentration solar cells*

Amonix has developed a back-junction, point-contact silicon solar cell which has many superior attributes for high-concentration system applications [1].

- High efficiency: The Amonix high-concentration silicon solar cell holds the world's record for performance for cells manufactured in a commercial environment (>26% efficiency at 300× concentration, 25°C).
- Ultraviolet light stability: No degradation in performance through system life.
- High power capability: Allows wide concentration application range (200×–500×) without significant loss in efficiency.

- Both electrodes on the same side - Allows for highly automated surface mount assembly methods.
- Industry standard manufacturing processes: Allows high-volume cell production specifically adaptable to large-volume computer chip foundries for low-cost manufacturing.

7.4.5 *The integrated high-concentration PV (IHCPV) system*

Development of the Amonix HCPV solar cell has paved the way for to commercial deployment of HCPV systems [2]. High-concentration PV systems offer several distinct advantages for low-cost power generation: (a) cost reduction through reduction in silicon area — a traditional silicon 10 cm diameter one-sun wafer with $75\,cm^2$ of active area produces 1 W while the same wafer produces the equivalent of 700 W at 500× concentration; (b) higher conversion cell efficiency at concentration vs. one-sun; and (c) inherently higher capacity factor compared to fixed-tilt, flat-plate systems in high direct normal insolation (DNI) areas because of its built-in tracking. Despite these intrinsic cost-reducing elements, high-concentration PV systems have not previously emerged for large-scale utility use because of (a) the earlier lack of a stable, high-performance, commercially available high-concentration solar cell, and (b) the high costs associated with PV modules, structure, tracking system, and ancillary equipment.

The Amonix solar cell now enables HCPV systems to be realized. However, to reduce costs, considerable effort has been applied so that all of the savings resulting from greatly reduced silicon usage does not get lost in the cost for the structure, tracker, and ancillary equipment required in the concentrating system. The result is the integrated high-concentration PV (IHCPV) array.

This new, innovative (patented) system concept eliminates much of the costly hardware used in earlier high-concentration designs. This has been accomplished by the simplification of the array structure which (1) eliminated earlier separate "box"-type modules mounted on structure assemblies, and substitutes an integrated design which combines both the load-bearing structure and the Fresnel lens/receiver plate elements, thus eliminating the need for separate modules, and (2) a novel, manufacture-worthy receiver plate which makes use of "circuit-board" construction techniques, with surface mount cell technology, eliminates costly and labor intensive cell packaging and interconnects. A schematic illustration of the IHCPV system is shown in Fig. 7.18.

Figure 7.19 shows a picture of an IHCPV system installation at a large utility site: Arizona Public Service Company's Solar Test and Research (STAR) facility.

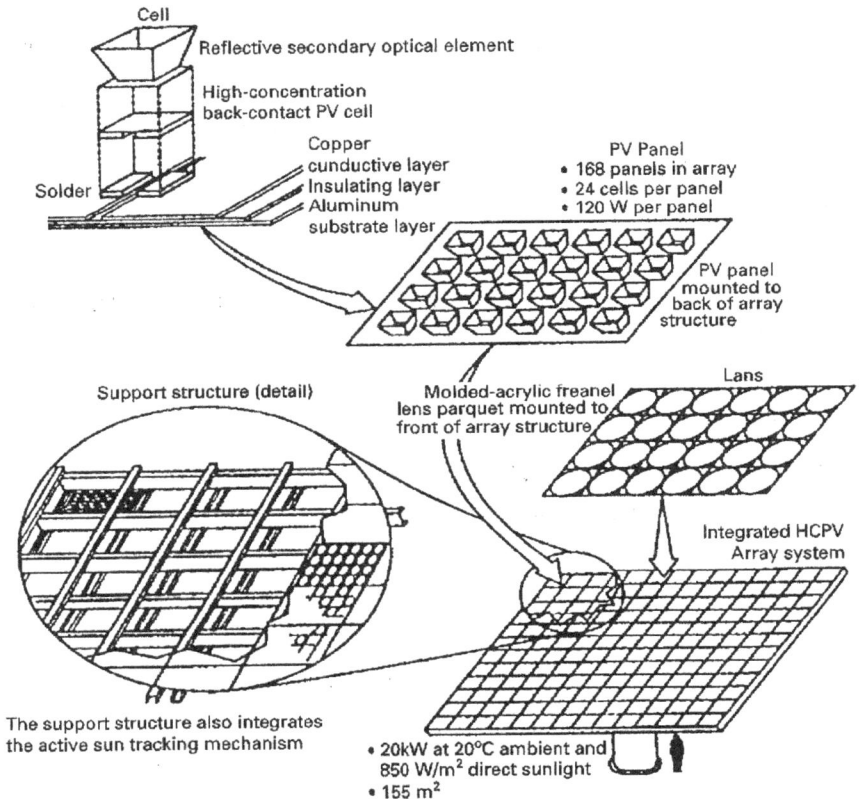

Figure 7.18:   Schematic of IHCPV system.

Figure 7.19:   Photograph of IHCPV System at the STAR facility.

Amonix has installed five full-scale 18 kW utility demonstration IHCPV systems which have demonstrated the maturity of the technology as a provider of cost-competitive utility-scale (multimegawatt) electric power. These systems have demonstrated 18% conversion efficiency at PVUSA conditions (more than twice the performance of other technologies). Now that the basic technology has been validated, attention can shift to economics of IHCPV as a low-cost near-term alternative PV system for large-scale solar electric power generation.

7.4.6   *Economics of IHCPV*

There are several factors, specific to IHCPV systems, which offer a near-term opportunity for low cost PV installations on a large scale:

- low cost of entry,
- enhanced energy production,
- reduced land utilization,
- accelerated benefits of volume production.

7.4.6.1   *Low cost of entry*

Billions of dollars have been spent on R&D and hundreds of millions on facilities for other PV technologies. It is estimated that each 10 MW increase in production capacity requires $20–25 million in capitalization. In contrast, Amonix uses a unique approach in order to adapt its complex solar technology (both cell and system) to commonly existing manufacturing infrastructures (e.g., high-volume semiconductor computer chip manufacturing facilities called foundries) rather than developing and capitalizing custom facilities around a specialized technology. Both the Amonix HCPV solar cell and the IHCPV system have been designed to standard industry manufacturing methods and processes. This enables IHCPV immediate high-volume market entry. This strategy has resulted in at least two major benefits:

(1) "Fast-Track" high-volume production capability. An enormous underutilized manufacturing capacity, equaling gigawatts of IHCPV, exists in the US alone. By using this semiconductor and other manufacturing capacity, IHCPV can achieve production levels of 50 MW/yr within 18 months at one-third the capital cost of other PVs. No other PV technology has this capability or capacity.

(2) Price/cost predictability. By using industry-standard high-volume manufacturing methods, Amonix can project costs with considerable authority and confidence.

High-volume production costs for semiconductor chips similar in complexity to the Amonix HCPV solar cell have been well established. Based on this, Amonix projects HCPV cell costs to be <20¢/W at the 50 MW/yr production level. Fresnel lens and structure cost projections are based on equally accurate information and analysis derived from existing manufacturing infrastructure. A conservative projected comparison of IHCPV vs. generic flat-plate system components in dollars per watt is shown in Table 7.3.

The figures are for installed costs at the dc level because dc to ac conversion costs are the same for both technologies. The projection

Table 7.3: IHCPV/ Flat-plate comparison.

	IHCPV	Flat-plate
Cell	$0.20	$1.20
Fresnel lens	$0.20	N/A
Receiver plate	$0.44	N/A
Modularization	N/A	$1.20
Subtotal	$0.89	$2.40
Structure	$0.45	$0.25
Tracker	$0.19	N/A
Installation	$0.20	$0.15
Total	$1.68	$2.80

Source: available literature

is for the year 1999 at 50 MWdc production rates per annum at a single factory.

7.4.6.2 *Enhanced energy production*

Sun-tracking PV systems such as concentrators actively track the sun maximizing energy production. At almost any installation site, tracking enhances energy production although the expenditure associated with tracking must always remain cost effective. Note that the cost comparison shown in Table 7.3 does not take increased energy production into consideration. Realistically, IHCPV produces up to 30% more electricity than competing technologies. A simple methodology for determining the benefits of enhanced energy production from HCPV systems has been reported [3]. By "normalizing" the dollar per watt comparisons between differing technologies using an energy production factor (EPF), a greater accuracy can be obtained relative to the value of competing PV concepts. The energy production factor is defined as

$$EPF = \frac{\text{Site \& tech. specific avg. daily solar radiation/Reference solar irradiance}}{\text{Fixed flat-plate site specific avg. daily radiation/1000 W/m}^2}$$

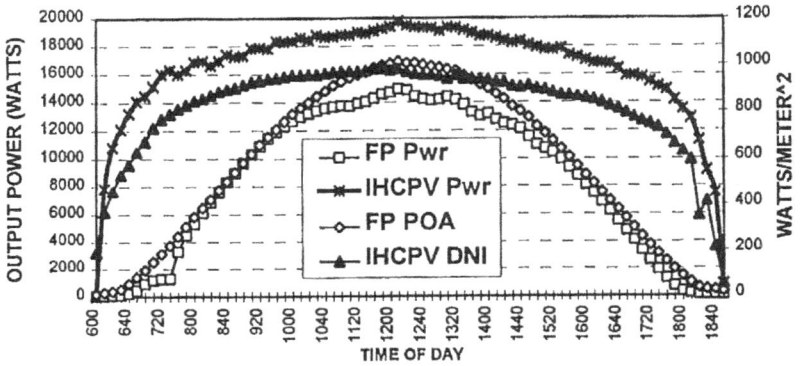

Figure 7.20:   Comparison of similarly sized fixed flat-plate and IHCPV systems showing greatly enhanced energy production.

Typical data taken for a side-by-side comparison of 20 kW peak systems graphically demonstrates that the IHCPV system produces considerably more electricity than a similarly sized fixed flat-plate system as shown in Fig. 7.20.

7.4.6.3   *Reduced land utilization*

IHCPV systems, due to their superior performance, require a significantly smaller land area than conventional PV, for generating equivalent amounts of electrical energy. This is an attractive benefit, especially in areas where real-estate costs can play a determining factor in power-plant location. With a system efficiency approaching 20%, the IHCPV system consumes one-fourth the area required by less efficient technologies such as thin-film. In addition to the obvious benefits, this implies for multi-megawatt installations, the small "footprint" of the IHCPV system is advantageous for distributed networking and adjunct installations.

7.4.6.4   *Accelerated benefits of volume production*

Perhaps the most significant aspect of IHCPV's advantages is its enhanced potential for meeting the cost thresholds needed for

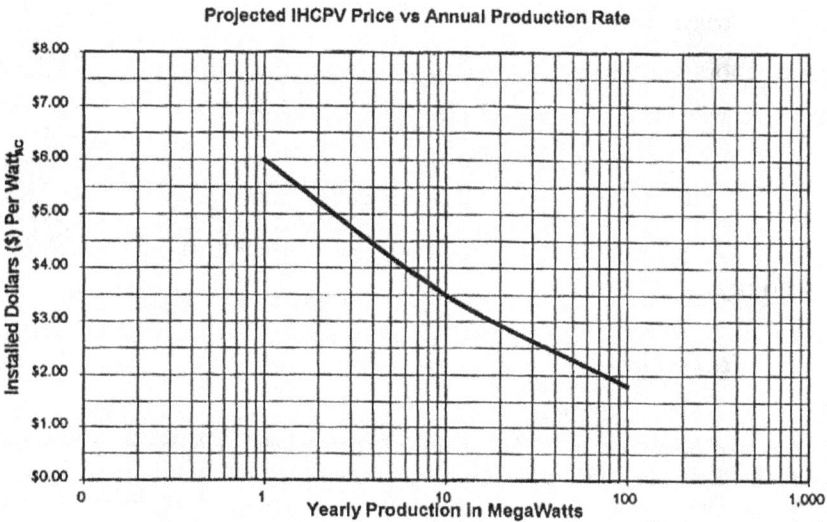

Figure 7.21:   Accelerated benefits of volume production.

large-scale deployment and utilization. No other PV technology is positioned to reap the benefits of volume production levels in an accelerated fashion as is IHCPV. By utilizing industry standard manufacturing facilities and methods as described above, IHCPV is capable of being produced today at large-volume levels without the multi-year "ramp-up" required by other technologies. And, because all of the elements of the IHCPV system (solar cell, lens, structure) were designed for volume manufacturing at the onset, IHCPV will transit to volume production much faster (Fig. 7.21).

7.4.7 *Summary*

A near-term, cost-effective alternative PV technology is available for large-scale deployment: integrated high-concentration PV (IHVPV). This system combines world record performance with low cost of market entry, enhanced energy production, reduced land utilization, and accelerated benefits of volume production.

7.4.8 *Acknowledgements*

The authors wish to acknowledge the sustaining support of Mr. Herb Hayden and Mr. Pete Eckert of Arizona Public Service Company, Dr. Edgar DeMeo of the Electric Power Research Institute, Mr. Ward Marshall of Central and South West Services, and Mr. Dale Green of Nevada Power Company. Their work, along with many others, has greatly accelerated the commercialization effort of the IHCPV technology.

7.4.9 *References*

[1] S. Yoon, V. Garboushian, "Commercialization of high-concentration back-junction point-contact photovoltaic cells; successful pilot production", *11th E.C. Photovoltaic Solar Energy Conference* , 1
[2] V. Garboushian, S. Yoon, G. Turner, A. Gunn, and D. Fair, "A novel high-concentration PV technology for cost-competitive utility bulk power generation", *IEEE 1st World Conf. on Photovoltaic Energy Conversion*, 1994, pp. 1060–1063.
[3] J. Allen Gunn, D. Fair, V. Garboushian, and S. Yoon, "A simple method for equitably comparing different types of PV systems based upon energy production", *13th European Photovoltaic Solar Energy Conference*, 1995, pp. 1804–1806.

7.5 An informal round-table discussion of Vahan Garboushian's text, led by David Faiman in the absence of the author

Arbib: Three years ago, here at Sede Boqer, Professor Kaneff made a persuasive case for using a single dish concentrator for irradiating an array of closely packed photovoltaic cells. Why does Garboushian prefer to illuminate each cell with its own concentrator, and with lenses rather than mirrors?

Faiman: I think there are several reasons. Chief among them are: (1) All of the optical alignment can be conveniently and accurately performed in the factory; (2) Transportation of their modules to the site and subsequent installation is much easier than for dish system components; (3) Cleaning of flat, sealed modules is much simpler than the corresponding operation for dishes.

Levy: One of the advantages of dish systems is that the array cooling subsystem produces readily packaged thermal energy that can be used. By contrast, Amonix's air-cooled modules waste all of that energy.

Faiman: I am sure that Vahan would not claim that dish systems do not have any advantages. The one you point out is relevant if and only if there is a need for thermal energy at the temperature at which, say, Kaneff's CPV system generates it. If there is no need for that energy then the advantage becomes a disadvantage in that the waste energy must be removed by some kind of heat exchanger, which adds extra cost to the system.

Medwed: If they would mount an array of standard solar cells on their dual axis tracker, they would be able to take advantage of the diffuse radiation which an expensive concentrating array wastes,

Faiman: Your observation might be correct if standard solar cells were an almost zero cost item. But at today's costs it ignores the economics. Capturing the diffuse radiation might increase the solar input by about 50%. However, the Amonix concentrator cells have almost twice the efficiency of standard silicon cells even without concentration. And their Fresnel lenses increase that advantage by a factor of several hundred. Their system looks extremely promising compared with standard PV arrays.

Grunbaum: Could they not increase their efficiency even further by using multijunction III–V cells?

Faiman: In principle, they might be able to do so, depending on the future commercial availability of stable cells at a cost and efficiency that warrant such a change-over. But for the time being, they have their own stable, high-efficiency silicon cells on hand, with all contacts on the cell's rear surface, and an industrial process for making connections and integrating them into a concentrator module.

In this regard, I have often felt that one of the great advantages of dish concentrators is that the cell module, which is a small part of the system, can easily be replaced if subsequent improvements in cell efficiency warrant a replacement. But I realize that Amonix could also exchange their rear panel with one that contains improved cells.

Weil: Doesn't the sudden on-off output of concentrator cells present a problem for their integration into a grid system?

Faiman: The output is only on-off for a single cell. A large module such as the Amonix IHCPV or Steve Kaneff's big dish would smooth this out to a certain extent, as the shadow of a cloud moves across it, and a large field of trackers or dishes would react to passing clouds in much the same way as a field of non-concentrator PV cells.

Chapter 8

1999

8.1 Editor's Foreword

For the final *Sede Boqer Symposium* of the 20th century, which took place in July 1999, we had the pleasure of three keynote speakers. One of them, Jacob S. Ishay, professor of physiology at Tel Aviv University, presented some fascinating experimental laboratory data he and his colleagues had observed on the effect of ultraviolet light on the ability of hornets to perform nest-digging activities followed by soil-dumping at distances seemingly too far from their nests for their body weight to allow. One of Ishay's speculations was that the hornets employ the photovoltaic effect in order to convert the ultraviolet radiation of sunlight into metabolic energy. This fascinating lecture was omitted from the present volume on the grounds that whether confirmed as the photovoltaic effect or not, the subject does not fall within the history of the development of solar power stations. Nevertheless, the interested reader is referred, as a useful starting point, to their journal publication: J. S. Ishay, V, Barenholz-Paniry and A. Bitler, "I. On hornet silk as a photodetector: considerations of current, voltage and resistance." *Physiological Chem. & Phys. & Medical NMR*, 29 (1997) 95–108.

On the other hand, our other two keynote speakers were ideally suited for rounding off the millennium. The late Peter Landsberg, whose deep physical/mathematical insights graced very many areas of quantum physics presented a broad-based review of the fundamentals of solar energy efficiency, shedding much light on the relationships among the various formulae (including his own) that have appeared over the years.

And in a step away from conventional solid state devices, Serdar Sariciftci presented the latest results about the plastic solar cells that he and his colleagues are developing that may, one day, enable the clothes we wear to generate the electricity that will be needed to power the various miniature digital devices that will enable us to maintain uninterrupted contact wherever we are, with one another and with data banks for all conceivable needs.

8.2 Theoretical bounds on solar cell efficiencies

A keynote lecture presented by Prof. Peter T. Landsberg, Faculty of Mathematical Studies, University of Southampton, Southampton, SO17 1BJ, UK.

8.2.1 *Contents*

- Introduction
- Thermodynamic efficiencies
- A simple "ultimate" photovoltaic efficiency
- General comments on efficiencies
- The effect of radiative recombination and of multiple sources
- The heterojunction cell with Auger effects and impact ionization
- Conclusion
- Appendix 8.A.1, 8.A.2, 8.A.3
- References

8.2.2 *Introduction*

I have been concerned with solar cells (among numerous other topics) since William Shockley awarded me a research contract on this topic in the 1970's. Since then I have been interested in many aspects of efficiencies and how they are limited by electron–hole recombination. This effect removes solar-generated particles which would otherwise contribute to the photocurrent. So it is a "bad thing", and therefore worthy of study on this and many other grounds. In fact, my concern with semiconductors goes back to before the first International Conference on Semiconductors organized by

H. K. Henisch in Reading, England, in 1950. The frontispiece in [1], which is rather compressed, shows Shockley, Brattain, Mott, R. W. Pohl and younger persons who were to make their mark later, like Charles McCombie and Oliver Heavens who became Professors of Physics in Reading and York respectively, and Dennis Sciama who became a distinguished Professor and cosmologist in Oxford.

A much better and larger copy of this picture, properly labelled, as well as some residue cheap copies of the book are available from me directly.

In the long period 1950–1999 I covered many questions in the solar cell area, lately in conjunction with collaborators. As the work reported here is based so heavily on it, I wish to thank them: Viorel Badescu, Pierre Baruch, John Liakos, Thomas Markvart and, alphabetically last, but longest in years, Alex de Vos.

I also recall some well-known persons with whom I had contact, but who are no longer with us. The famous American trio William Shockley, Walter Brattain and John Bardeen, Walter Schottky — a real German gentleman, and the Russian colleagues B. Wul and S. G. Kalashnikov, who were much in evidence at the World Exhibition held in Brussels in 1956 [2]. Figure 8.1 gives perhaps a flavor of this occasion. Then there were Walter Bloss and our own Neville Mott and Trevor Moss whom we lost more recently. Among those of this generation (more or less!) who fortunately survive, and whose semiconductor interests overlapped happily with mine from time to time, are Karl Boer, Heinz Henisch, M. H. Pilkuhn, Hans Queisser, Alexander Rogachev, Jan Tauc and Peter Würfel. Then there are my former, but still active, doctoral students in the semiconductor area Mike Adams, "Slim" Beattie, Ashok Pimpale and Eckehard Schöll. There are other I should mention, as such lists are always incomplete, but this will have to do. I thank them all for their work, their help and their support. Particularly thanks are due to David Faiman and The Academic Study Group on Israel and the Middle East for enabling me to make this (my second) visit to Israel. I am also indebted to Dr. V. Badescu and Dr. S. Zh. Karazhanov, Tashkent, for recent discussions.

Figure 8.1:   Solid State Physics was represented at the Brussels World Exhibition in 1958. "Le Belgique Joyeuse" was part of it where this photograph was taken. From left to right: Manuel Cardona (later Professor in Stuttgart), Bill Paul (later Professor at Harvard), Peter Landsberg, the late Trevor Moss, Dennis Smith (then Trevoi's student and later Professor in Edinburgh and Fellow of the Royal Society).

8.2.3   *Thermodynamic efficiencies*

When asked about upper limits of an energy conversion process, one thinks of Sadi Carnot and his successors. Thus, if T_s and T_p are the absolute temperatures of the surroundings (the "ambient") and of the pump respectively, we have at least four thermodynamic formulae which are used in our work (numerical values are given for $T_s = 300\,\mathrm{K}$, $T_p = 6000\,\mathrm{K}$). They are displayed in Fig. 8.2 and are

$$\eta_C = 1 - T_s/T_p = 19/20 = 95\% \tag{8.1}$$

$$\eta_{CA} = 1 - (T_s/T_p)^{1/2} = 77.6\% \tag{8.2}$$

$$\eta_{PPL} = 1 - 4T_s/3T_p + (T_s/T_p)^4/3 = 93.3\% \tag{8.3}$$

$$\eta_{pt} = [1 - (T_c/T_p)^4](1 - T_s/T_c) = 85\% \tag{8.4}$$

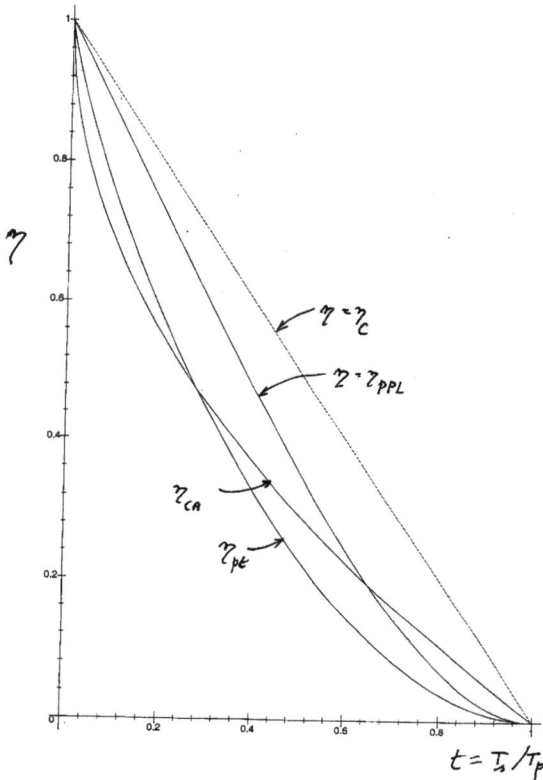

Figure 8.2: The efficiencies (8.2) to (8.4) as a function of T_s/T_p.

where T_c is determined by T_s and T_p via the quintic equation

$$4T_c^5 - 3T_s T_c^4 - T_s T_p^4 = 0 \qquad (8.5)$$

As clear from Fig. 8.2, the efficiencies reach the maximal values (unity) for $T_s = 0$ and by Equation (8.5) it also requires $T_c = 0$. Low temperature surroundings clearly help one to find good efficiency values. In contrast, low pump temperatures, such as $T_p = T_s = T_c$, reduce the efficiency to zero as one has effectively an equilibrium situation.

These four results apply best to photothermal conversion as no energy gap has been introduced. This is of course needed for

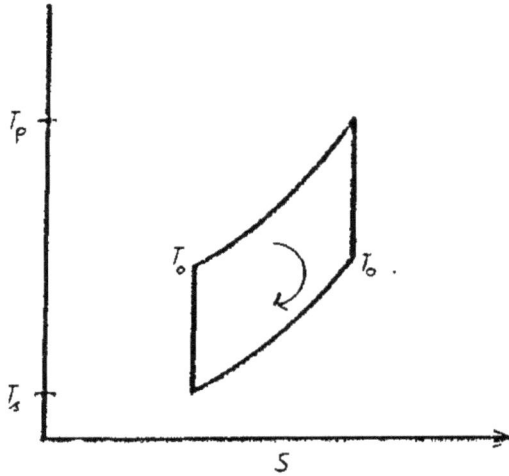

Figure 8.3:   A thermodynamic cycle to obtain Equation (8.2).

photovoltaic conversion. These four formulae are useful because they involve only two or three temperatures, and they will now be deduced. Unfortunately, actual experimental efficiencies lie well below these values, at most at about 30% for normal conditions.

The suffix CA stands for Curzon–Ahlborn, who discovered the formula (8.2) independently of the original earlier authors [3, 4]. One can obtain it [5] by using a fluid of constant-heat capacity C which executes a reversible cycle. It is heated from T_0 to T_p (see Fig. 8.3), using an infinity of heat reservoirs.

Adiabatic cooling to T_0 follows. Then there is rejection of heat done reversibly, using another sequence of heat reservoirs so that the fluid reaches T_s. Adiabatic heating to T_0 then closes the cycle. Entropy conservation gives

$$\int_{T_o}^{T_p} \left(\frac{C}{T}\right) dT = \int_{T_s}^{T_o} \left(\frac{C}{T}\right) dT \qquad (8.6)$$

that is, $T_p/T_0 = T_0/T_s$. The efficiency of work production is

$$\eta_{CA} = [C(T_p - T_0) - C(T_0 - T_s)]/[C(T_p - T_0)]$$
$$= 1 - [(T_p T_s)^{1/2} - T_s]/[T_p - (T_p T_s)^{1/2}]$$

This is

$$1 - [T_s^{1/2}(T_p^{1/2} - T_s^{1/2})]/[T_p^{1/2}(T_p^{1/2} - T_s^{1/2})]$$

and yields the result (8.2). It shows that the theoretical reversible efficiency of an engine can be considerably reduced with the desirable effect of bringing it closer to experimental results.

The original derivations of the result (8.2) [4] assumed maximum power conditions and linear heat transfer laws which led to the introduction of intermediate temperatures. This type of engine, like the Carnot engine, can act as a thermodynamic model for the conversion of solar energy into work. In fact the result (8.2) holds, at least approximately, for a whole class of thermodynamic engines [5].

On balancing energy fluxes (Φ) and entropy fluxes (Ψ) one obtains efficiencies of the type (8.3), see Fig. 8.4.

Let two suffices pc, cs, etc., denote transfers from pump to converter, from converter to the surroundings, etc. and let $\dot{Q}$ and $\dot{W}$ be heat and work fluxes emitted by the converter. Then the energy and entropy fluxes received by the converter are for a steady state

$$\Phi_{pc} + \Phi_{sc} = \Phi_{cp} + \Phi_{cs} + \dot{Q} + \dot{W} \tag{8.7}$$

$$\Psi_{pc} + \Psi_{sc} = \Psi_{cp} + \Psi_{cs} + \dot{Q}/T_x - \dot{S}_g \tag{8.8}$$

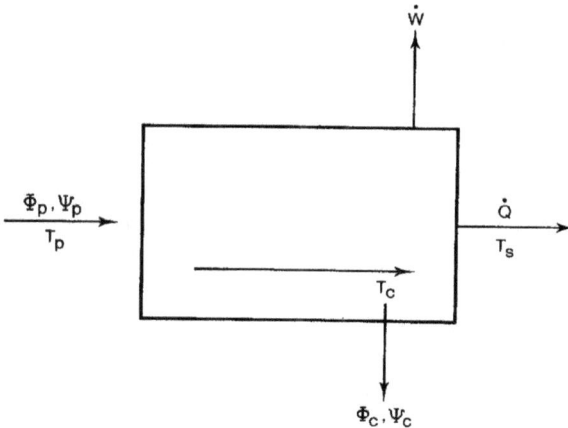

Figure 8.4: A general model of energy conversion. It can be treated quantitatively by the use of balance equations and then leads to the efficiencies given in Section 8.2.4

T_x denotes the temperature at which the cell emits the heat flux $\dot{Q}$ through its cover plate.

In general T_x is expected to be smaller than T_c. Eliminating $\dot{Q}$ between (8.7) and (8.8),

$$\eta \equiv \dot{W}/\Phi_{pc} = 1 - T_x\Psi_{pc}/\Phi_{pc} + (T_x\Psi_{cs}/\Phi_{cs} - 1)\Phi_{cs}/\Phi_{pc} + B \quad (8.9)$$

where

$$B \equiv (\Phi_{sc} - \Phi_{cp} + T_x\Psi_{cp} - T_x\dot{S}_g)/\Phi_{pc} \quad (8.10)$$

The sc and cp terms are expected to be negligible, so that B is negative in virtue of the entropy generation flux $\dot{S}_g$. Neglecting B, gives therefore an inequality [6]:

$$\eta < \eta_l(T_x) \equiv 1 - T_x\Psi_{pc}/\Phi_{pc} + (T_x\Psi_{cs}/\Phi_{cs} - 1)\Phi_{cs}/\Phi_{pc} \quad (8.11)$$

$$\sim 1 - 4T_x/3T_p + (4T_x/3T_c - 1)(T_c/T_p)^4) \quad (8.12)$$

For (8.12) it has been assumed that the radiation exchanges are restricted to black-body radiation. This is a versatile formula because it covers many important special cases. But it is not very useful by itself: we find as a special case the so-called Petela–Press–Landsberg efficiency (8.3) [7], provided T_x is taken as $T_c = T_s$. The surroundings, the converter and the cover plate are then all at the same temperature.

The Carnot efficiency (8.1) is found for

$$T_x = 3T_c/4 = 3T_s/4 \quad (8.13)$$

We see from this, as expected, that the lower T_x gives the higher efficiency. Thus, (8.12) is a kind of master formula which includes (8.1) and (8.3) as special cases. It can be forced to yield (8.2) as well, but the substitution required for T_x, namely

$$4T_x/3T_p = [(T_s/T_p)^{1/2} - (T_c/T_p)^4]/[1 - (T_c/T_p)^3] \quad (8.14)$$

has no obvious physical interpretation.

In the real photothermal case (8.4), one heats a radiator R with the incident (solar) energy. Some of the re-radiated energy flux, $\dot{Q}'$ acts as a heat source for a Carnot engine, the other part goes to

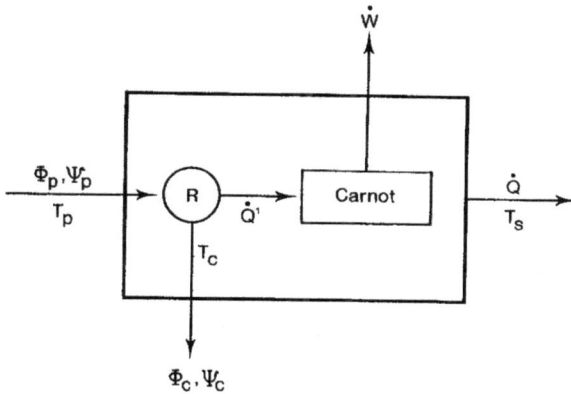

Figure 8.5: A simple photothermal conversion device.

the converter at temperature T_c. The work flux $\dot{W}$ comes from the Carnot engine (Fig. 8.5).

The efficiency of the device is

$$\eta_{pt} = \dot{W}/\Phi_p = (\dot{Q'}/\Phi_p)(\dot{W}/\dot{Q'}) = [(\Phi_p - \Phi_c)/\Phi_p](1 - T_s/T_c) \quad (8.15)$$

For black-body radiation one finds the result (8.4). Its maximum with respect to T_c satisfies (8.5). Thus, we have found our four (very high) efficiencies. Other theoretical efficiencies have been proposed. For example, Jeter in his Table 1 [8] gives (8.1) as his fourth version, and a modified form of (8.3) as his number 3. One form of (8.3) is discussed extensively by Kabelac [9] and a generalized form which allows for different solar concentrations appears as equation (3.5) in [10]. Attempts at unifying these results also exist [11].

8.2.4 *A simple ("ultimate") photovoltaic efficiency*

Specifically photovoltaic efficiencies are obtained by introducing an energy gap E_g across which incident photons can excite the electrons of a semiconductor. They leave behind holes, and electrons and holes are then separated in space by the internal electric field (as in a p–n junction). This yields a photovoltage and a photocurrent.

A simple argument for the maximum efficiency (46,500 Suns) of a photovoltaic converter uses as a key idea that a photon flux can be used to calculate energy output. Let $y = h\nu/kT_p$, where T_p is again the pump temperature, for example, the temperature of solar radiation received on earth. Let $g(y)\ dy$ be the number of photon states between y and $y + dy$, multiplied by a constant so that $g(y)$ represents a number flux. Normally $g(y) \propto y^2$. Let y_g be the energy gap E_g divided by kT_p. Further, let

$$f(y) = [\exp(y) - 1]^{-1}$$

be the equilibrium photon number in the radiation mode y. Then we assume (i) that the pump surrounds the cell (which corresponds to maximum concentration of the incident solar energy), and (ii) that each absorbed photon contributes the energy gap to the photovoltaic energy output. The energy flux produced by the device is then E_g multiplied by the number flux of photons. For an efficiency this quantity has to be divided by the incident energy flux. The so-called "ultimate" efficiency of the device is then [12, 13]

$$\eta_{ult}(y_g) \equiv \left[y_g \int_{y_g}^{\infty} g(y)f(y)dy \right] \Big/ D \tag{8.16}$$

where

$$D \equiv \int_{0}^{\infty} yg(y)f(y)dy$$

is the input flux Φ_p divided by kT_p:

$$D = \Phi_p/kT_p.$$

The integrand $g(y)f(y)$ is illustrated in Fig. 8.6 (showing a maximum) for $g(y) = y^2$.

Since $\eta_{ult}(0) = 0$ and $\eta_{ult}(\infty) = 0$, a maximum ultimate efficiency exists, at $y_g = y_{g0}$, say, such that $0 < y_{g0} < +\infty$. One finds by differentiating (8.16) with respect to y_g that y_{g0} must so divide the function $g(y)f(y)$ into two parts, that the rectangle of area $y_{g0}g(y_{g0})f(y_{g0})$ to the left of $y_g = y_{g0}$ has the same area as is found under the curve to the right of y_{g0}, that is,

$$y_{g0}g(y_{g0})f(y_{g0}) = \int_{y_{g0}}^{\infty} g(y)f(y)dy. \tag{8.17}$$

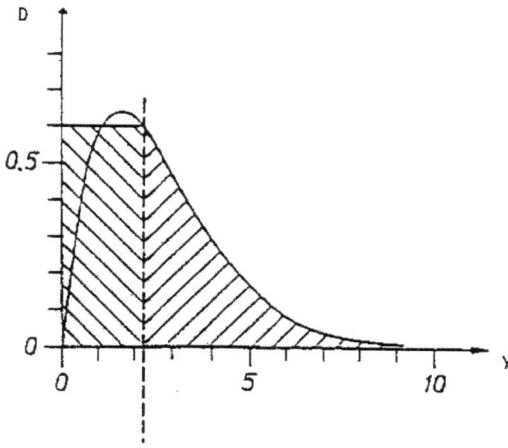

Figure 8.6:   The integrand of Equation (8.16).

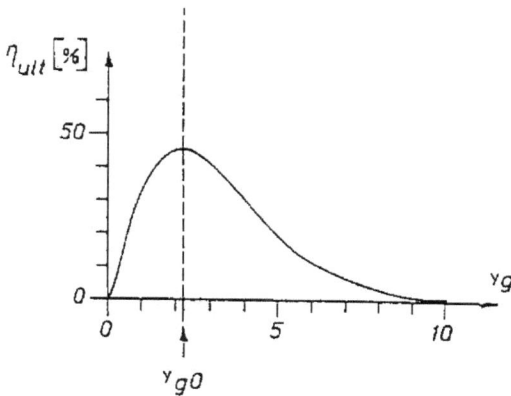

Figure 8.7:   The ultimate efficiency (8.16).

The curve of $\eta_{ult}(y_g)$ (Fig. 8.7) shows a well-known maximum at 44% near $y_{g0} = 2.2$.

For a concentration of merely one sun this maximum theoretical efficiency is reduced to about 30%. We note that as far as photovoltaic conversion is concerned this most specific of our models so far yields the lowest efficiencies. They are still too high to be attained but they are not as unrealistic as the efficiencies (8.1)–(8.4).

More realistic theories are of course available which give lower theoretical efficiencies. Some are discussed in Section 8.2.6.

8.2.5   *General comments on efficiencies*

The overestimate of the efficiency just obtained (in Section 8.2.4) is due to the neglect of radiative and non-radiative recombination. This leads to the loss of electron–hole pairs which can therefore not contribute to the current. This effect is due to electronic energy being given to the lattice (electron phonon collisions) and energy transfer to other electrons by the Auger effect (electron–electron collisions). There is also the effect of band gap shrinkage.

On the other hand, even higher efficiencies have been envisaged by having several materials with decreasing energy gap in series. These are the tandem cells with experimentally so far at most four or five in series. But the theory for an infinite number has also been worked out by de Vos and his colleagues [7], as well as by John Parrott and others. This procedure makes better use of the high energy photons: for, if they merely give up the gap energy to create one electron–hole pair, then the remaining energy would go to heat up the lattice which is not an efficient use of the available energy.

Another way of imagining a use for the higher energy photons is to let them produce more than one electron–hole pair. This "impact ionisation" [14] takes place in a cell in any case to a greater or lesser degree and it leads to a modification of Equation (8.16), as shown in Appendix 8.A.1. Theoretical studies of it can be used to indicate the maximum attainable efficiency if this phenomenon is fully exploited. They can also be used to indicate band structures which favor the occurrence of impact ionization [15]. Additional electron hole pairs can also be produced by photon reabsorption, sometimes known as "photon recycling".

These two extreme cases (the infinite tandem cell and unlimited impact ionization) lead remarkably to similar high efficiencies; for one sun the maximum efficiencies are of the order 61–68% and they are about 86–88% for maximal solar concentration. For an introduction to the theory and a recent list of about 100 theoretical and experimental efficiencies, see [16].

The experimental results are the lower ones in this list. Even for the favorable monochromatic irradiation the efficiency drops already to 45%. For a two-stage GaAs/GaSb tandem cell with concentration of $C = 50$ suns one can achieve the very respectable 34% (still only about half the theoretical efficiencies quoted above). The value for an InP/GaInAs cell is about 32%. One can achieve 30% for concentration $C = 100$. For a small-area GaAs/GaInAsP tandem cell at $C = 39.5$ one has found 30%. For unconcentrated radiation one can reach 25% for a GaInP/GaAs/Ge tandem cell. If one uses a monogap crystalline Si solar cell and concentrated radiation, one can reach "only" 23% and 24% if GaAs is used instead. For references, see [16].

8.2.6 The effect of radiative recombination and of multiple sources

In order to take account of extra sources ($i = 1, 2, \ldots$) of radiation one has to include a geometrical factor Γ_i for each source. For sources covering a small part of the sky, this is approximately proportional to the solid angle subtended by the source at the earth multiplied by the cosine of the angle of incidence of the light on the solar cell. Normalized for a hemisphere, this means that the Γ_i's add up to unity in important cases, see appendix 1 in [17]. The current density becomes for n blackbody sources of temperatures $T_1, T_2, \ldots, T_n$

$$I(V) = \int_{\theta_{yg}}^{\infty} \left[\sum_{i=1}^{n} \frac{\Gamma_i}{e^{yt_i} - 1} - \frac{\gamma}{e^{y-v} - 1} \right] g(y) dy \qquad (8.18)$$

where $g(y)$ is discussed in Section 8.2.4. The sum represents the n generating terms of the type (16) with y now E/kT_s and $t_i \equiv T_s/T_i$. The last term allows for re-radiation from the converter through radiative recombination at voltage V, γ being another geometrical factor. It takes into account the surface areas and solid angles which participate in the radiation from the converter. This reduction in the current produced must be included in any reasonable theory.

The "$-v$" which occurs in the denominator is due to the fact that the distribution of radiation in a non-equilibrium semiconductor ceases to be black-body by "growing" a kind of chemical potential $kT_s v \equiv q(F_e - F_h)$ given by the difference in quasi-Fermi levels [18] and called "the photon chemical potential".

Any improved theory based on (8.18) is expected to give lower maximum theoretical efficiencies than those described in this paper since it is more realistic.

Let us look for the optimum energy gap by differentiating the work flux $VI(V)$ with respect to y_g at constant V. One should also differentiate at constant E_g with respect to V, but we shall neglect this, as it is unlikely to make much difference [19]. Our result will therefore be approximate. The maximization arising from $(\partial I(V)/\partial y_g)_V = 0$ is easily seen to yield an approximate optimal value of V, given E_g, or an approximate optimal value of E_g, given V:

$$\sum_{i=1}^{n} \frac{\Gamma_i}{e^{E_g/kT_i} - 1} = \frac{\gamma}{e^{(E_g - qV_{app})/kT_s} - 1} \tag{8.19}$$

or

$$\sum_{i=1}^{n} \frac{\Gamma_i}{e^{E_{g,app}/kT_i} - 1} = \frac{\gamma}{e^{(E_{g,app} - qV)/kT_s} - 1} \tag{8.20}$$

Let us introduce

$$z(E_g) \equiv \ln \left\{ 1 + \frac{\gamma}{\sum_{i=1}^{n} \Gamma_i/(e^{\frac{E_g}{kT_i}} - 1)} \right\} \tag{8.21}$$

Then (8.19) and (8.20) are

$$qV_{app} = E_g - kT_s z(E_g), \tag{8.22}$$

$$E_{g,app} = qV + kT_s z(E_{g,app}). \tag{8.23}$$

A striking special case arises for one source (of temperature T_p) surrounding the converter, in what is called a 2π-geometry. In that case (8.21) yields, with $T_1 \equiv T_p$ and for $\Gamma_1 = \gamma$, the simple result

$$z(E_g) = E_g/kT_p$$

Thus, (8.22) furnishes the Carnot factor in that case

$$qV_{\text{app}} = E_g(1 - T_s/T_p).$$

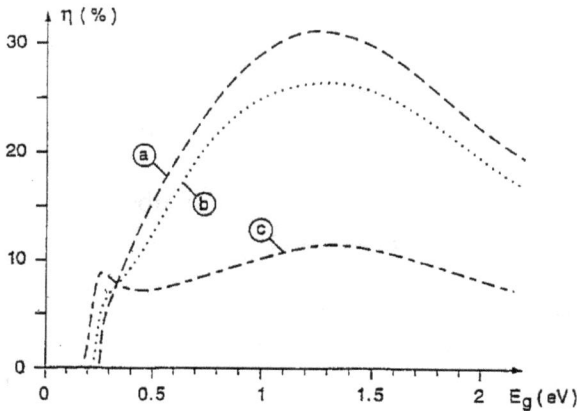

Figure 8.8: Efficiency of a solar cell subject to the effect of the surroundings and of sources 1 and 2. $T_p \equiv T_1 = 6000\,\mathrm{K}$, $T_2 \equiv T_3 = 300\,\mathrm{K}$, $\Gamma_p \equiv \Gamma_1 = 2 \times 10^{-5}$, $\Gamma_s \equiv \Gamma_3 = 1 - \Gamma_1 - \Gamma_2$. (a) $\Gamma_2 = 0.001$, $C = 50$, $d = 7.07$. (b) $\Gamma_2 = 0.01$, $C = 500$, $d = 22.36$. (c) $\Gamma_2 = 0.1$, $C = 5000$, $d = 70.71$.

See also [20] for a simple discussion of the occurrence of this factor.

More details and a specimen curve of an efficiency if two sources are present can be found in [17]. We here reproduce this curve as Fig. 8.8.

It gives the efficiency defined as the output power divided by the sum of the input powers from sources 1 and 2. As before C denotes the light concentration factor assumed. The symbol d gives the effective diameter of source 2, and is given by $(\Gamma_2/\Gamma_p)^{1/2}$ multiplied by the apparent diameter of the sun.

By micromachining of the surface the outgoing radiation can be confined to a small solid angle leading to higher values of $I(V)$ in (8.18), and hence to higher efficiencies, by making y very small. This proposal has been pioneered by the Spanish school of photovoltaics [21, 22]. For their more recent work see [23, 24].

8.2.7   *The heterojunction cell with Auger effects*

With two cells in tandem one expects an improved performance; however, the best efficiencies for unconcentrated radiation are in the 25–30% region for both monogap and tandem cells. The results depend on many factors, notably on the cell size. The greater the area, the harder it is to attain a high efficiency because of the increased chance of including some defective material.

We will not here give the theory for heterojunction cells. The main idea is that the higher gap cell is met first by the radiation. Some of the radiation transmitted is then absorbed and converted in subsequent units. In this process one may also take account of impact ionization. This can be done in various ways; in our work we always insisted [14] on using a probability $P(E)$ that an electron which has enough energy E to impact ionize, will actually do so. This introduces unfortunately yet another parameter, but a simple theory serves as a guide as to its value [25] as explained in Appendix 8.A.2 and Fig. 8.A.2. This probability depends on photon energy and we find that the efficiency of a cell for $P = 1$ is lowered by of the order of 5% if Auger recombination is included, but only by 3% if Auger recombination and impact ionization are both included. The point is this: in radiative studies one should include with radiative emission also absorption since emission and absorption are related by detailed balance in equilibrium. Similarly, one should include with impact ionization also Auger recombination since they, too, are related by detailed balance in thermal equilibrium [1, p. 394]. Our results [26, 27] indicate the order of magnitude of the error made if in theoretical considerations Auger effects are neglected and impact ionisation is treated with in effect $P = 1$. This latter has in fact been a usual procedure in the literature [15, 28].

Many variables and parameters are involved in this work [26, 27]. Four of them are the energy threshold parameters for Auger recombination, since the partner electron making a transition between bands cannot do so from the top of the valence band owing to the constraints of energy and momentum conservation. In fact, the participating particles have to occupy states away from the band edges in order to participate, and this reduces their probability of occurrence, and hence the probability of the whole process. This is illustrated in Fig. 8.9 and Appendix 8.A.3.

There are two such parameters for electrons (one in each semiconductor) and two for holes. Then there are minority carrier lifetimes,

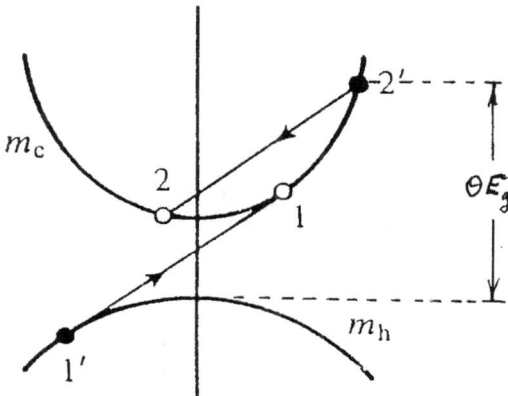

Figure 8.9:   The diagram illustrates the threshold energy for impact ionization in parabolic bands.

band–band radiative and Auger recombination coefficients as well as the usual diffusion coefficients, effective densities of states, etc.

One broad conclusion from our results is the following: Band–band Auger effects shift the optimum energy gaps of both heterojunction materials to higher values, while impact ionization has the opposite effect. The model also confirms one's expectation that the best maximum efficiencies are obtained for good impact ionization, low radiative and Auger recombination, and thin active layers.

8.2.8 *Conclusion*

Among recent, but still not fully confirmed, ideas to increase the efficiencies of solar cells are the following:

(a) It has been suggested that in equilibrium there may be as many as 10^{17} cm^{-3} excitons [29] in room temperature silicon. If they are split up in the field of a p–n junction, they can produce extra holes and electrons. This could have the doubly beneficial effect of decreasing the dark saturation current and increasing the light-generated current in a solar cell [30].

(b) The insertion of impurities placed appropriately on the energy band scale could enable better use to be made of the lower-than-gap-energy photons contained in the solar irradiation [31, 32].

(c) It would be very important if Auger recombination could be reduced or avoided [33]. One suggestion is to so arrange one's band gap engineering (quantum confinement, effect of strain, superlattice formation) that the threshold energies of which a simple example is given in Fig. 8.9, can be made large. They result from the need to satisfy energy and momentum among the electronic states simultaneously. For parabolic bands which can be described by effective masses one can consider the matter by inspecting the threshold energies tabulated for five band structures in [1, p. 250] and consider ways of making them large. Non-parabolic bands pose, of course, this problem in a different form. By strain-splitting of bands and by quantum confinement one can also make inaccessible states which can easily satisfy energy and momentum conservation, thus making Auger recombination harder: a desirable result.

It may thus be hoped that as new ideas enter this field of applied science higher efficiencies, and hence more widely economic solar cells, will be produced. For those of us who have helped in the struggles of engineers and scientists over the last 40 years for each fraction of a percent improvement, this will be a great reward. If wide domestic use became economic, it would be more than a reward, it would be a blessing for mankind.

8.2.9   *References*

[1] P.T. Landsberg *Recombination in Semiconductors* (Cambridge, University Press, 1991).

[2] M. Desirant (ed.) *Solid State Physics in Electronics and Telecommunications*, two volumes, (London, Academic Press, 1958).

[3] I.I. Novikov "The efficiency of atomic power stations" *J. Nuclear Energy II* **7** (1958) 125, translated from *At. Ener.* **3** (1957) 409, according to A. Bejan *Entropy Generation and Minimization* (Boca Raton, CRC Press, 1995), p. 208.

[4] F. Curzon and B. Ahlborn "Efficiency of a Carnot engine at maximum power output" *Am J Phys* **43** (1975) 22–24.

[5] P.T. Landsberg and H.S. Leff "Thermodynamic cycles with nearly universal maximum-work efficiencies" *J Phys A* **22** (1989) 4019–4026.

[6] P.T. Landsberg and V. Badescu "Efficiency and thermodynamic analysis of a two-temperature solar unit" *2nd World Conference on Photovoltaic Energy Conversion* Vienna, July 1998, p. 62.

[7] J.E. Parrott "The limiting efficiency of an edge-illuminated multigap solar cell" *J Phys D* **12** (1979) 441–450.
A de Vos *Endoreversible Thermodynamics of Solar Energy Conversion* (Oxford, University Press, 1992), and papers cited there.

[8] S.M. Jeter "Maximum conversion efficiency for the utilization of direct solar radiation" *Solar Energy* **26** (1981) 231-236. Comment by M. Castañs in *Solar Energy* **30** (1983) 293.

[9] S. Kabelac "A new look at the maximum conversion efficiency of black-body radiation" *Solar Energy* **46** (1991) 231-236. Comment by V. Badescu in *Solar Energy* **50** (1993) 379.

[10] P.T. Landsberg and P. Baruch "The thermodynamics of the conversion of radiation energy for photovoltaics" *J Phys A* **22** (1989) 1911–1926.

[11] A. Bejan "Unification of three different theories concerning the ideal conversion of enclosed radiation" *J Solar Energy Engineering* **109** (1987) 46-51. Comment by V. Badescu in *ibid* **110** (1988) 349–350.

[12] D. Trivich and P. Flinn "Maximum efficiency of solar energy conversion by quantum processes" in *Solar Energy Research* F. Daniels and J. Duffie eds, (London, Thames and Hudson, 1955), p. 143.

[13] W. Shockley and H.J. Queisser "Detailed Balance limit of efficiency of p-n junction solar cells" *J App Phys* **32** (1961) 510–519.

[14] P.T. Landsberg, H. Nussbaumer and G. Willeke "Impact ionization and solar cell efficiency" *J App Phys* **74** (1993) 1451–1452.

[15] M. Wolf, R. Brendel, J.H. Werner and H.J. Queisser, Solar cell efficiency and carrier multiplication in $Si_{1-x}Ge_x$ alloys" *J App Phys* **83** (1998) 4213–4221.

[16] P.T. Landsberg and V. Badescu "Solar energy conversion: List of efficiencies and some theoretical considerations, Parts I and II" *Prog Quantum Electronics* **22** (1998) 211–230 and 231–255.

[17] P.T. Landsberg, A. de Vos, P. Baruch and J.E. Parrott "Multiple source photovoltaics" in *Entropy and Entropy Generation* J.S. Shiner ed (Kluwer Academic Publishers, 1996) pp. 175–195.

[18] A one-page non-specialist survey by the present author appears in a birthday issue for Claude Benoit à Ia Guillaume, *Annals de Physique, Colloque 2, Supplement au no 3*, vol. 20, June 1995, page C2-393, entitled "The photon chemical potential".

[19] P. Baruch, P.T. Landsberg and A de Vos "Thermodynamic limits to solar cell efficiencies for illumination conditions" *11th European photovoltaic solar energy conference, Montreux*, October 1992, p. 283.

[20] P.T. Landsberg and T. Markvart "The Carnot Factor in Solar Cell Theory" *Solid-State Electronics* **42** (1998) 657–659.

[21] G.L. Araujo and A. Marti "Generalized detailed balance theory to calculate the maximum efficiency of solar cells" *11th European photovoltaic solar energy conference* Montreux, October 1992, p. 142.

[22] G.L. Araujo and A Marti "Absolute limiting efficiencies for photovoltaic energy conversion" *Solar Energy Materials and Solar Cells* **33** (1994) 213–240.

[23] A. Marti, J.L. Balenzategui and R.F. Reyna "Photon recycling and Shockley's diode equation" *J App Phys* **82** (1997) 4067–4075.

[24] A. Luque and A. Marti "Increasing the efficiency of ideal solar cells by photon induced transitions at intermediate levels" *Phys Rev Lett* **78** (1997) 5014–5017.

[25] J.K. Liakos and P.T. Landsberg "A simple model calculation of impact ionization probabilities and its consequences for solar cell efficiencies" *13th European Photovoltaic Solar Energy Conference* Nice, October 1995, p. 1235.

[26] H. Kiess and W. Rehwald "On the ultimate efficiency of solar cells" *Solar Energy Materials and Solar Cells* **38** (1995) 45–55.

[27] J.K. Liakos and P.T. Landsberg "Auger recombination and impact ionization in heterojunction photovoltaic cells" *Semicond. Sci. and Technol.* **11** (1996) 1895–1900.

[28] J.H.Werner, S. Kolodinski and H.J. Queisser "Novel optimization principles and efficiency limits for semiconductor solar cells" *Phys Rev Lett* **72** (1994) 3851–3854.

[29] D.E. Kane and R.M. Swanson "Effect of excitons on apparent band gap narrowing and transport in semiconductors" *J App Phys* **73** (1993) 1193–1197.

[30] R. Corkish, D. S.-P. Chan and M.A. Green "Excitons in silicon diodes and solar cells: A three-particle theory" *J App Phys* **79** (1996) 195–203.

[31] M.J. Keevers and M.A. Green "Efficiency improvements of silicon solar cells by the impurity photovoltaic effect" *J App Phys* **75** (1994) 4022–4031.

[32] H. Kasai and H. Matsumura "Study for improvement of solar cell efficiency by impurity photovoltaic effect" *Solar Energy Materials and Solar Cells* **48** (1997) 93–100.

[33] C.R. Pidgeon, C.M. Ciesla and B.N. Murdin "Suppression of non-radiative processes in semiconductor mid-infrared emitters and detectors" *Prog. in Quantum Electronics* **21** (1997) 361–419.

Appendix 8.A.1: Amendment of equation (8.16) by impact ionization

How does one amend the simple efficiency formula (8.16) to allow for impact ionization? One would expect an extra term in the numerator to take account of the additional energy flux arising from this new effect:

$$\eta(y_g) = \frac{1}{D} \left(y_g \int_{y_g}^{\infty} g(y)f(y)dy + I \right)$$

The form of I is in the simplest case [14]

$$I = P y_g \int_{\theta y_g}^{\infty} g(y) f(y) dy$$

Here, P is an average probability $P(E)$ of impact ionization (introduced in Section 8.2.7). It is multiplied by the photon flux arising from impact ionization. This is further multiplied by the (reduced) energy E_g contributed by each participating photon. The simplest form of the result is

$$\eta(y_g) = \frac{y_g}{D} \left(\int_{y_g}^{\infty} g(y) f(y) dy + P \int_{\theta y_g}^{\infty} g(y) f(y) dy \right)$$

This can of course be extended from a homojunction to a heterojunction; see [26, 27], where a variety of numerical results are given.

Appendix 8.A.2: Probability of impact ionization $P(E)$: A rough estimate

Suppose a photon of energy somewhat in excess of twice the band gap energy E_g creates an electron of energy E in the conduction band, as shown in Fig. 8.A.2. Of the energy losing transitions 1, 2, 3, etc., which it can make, only transitions of type 3 deliver enough energy to raise an electron from the valence band to the conduction

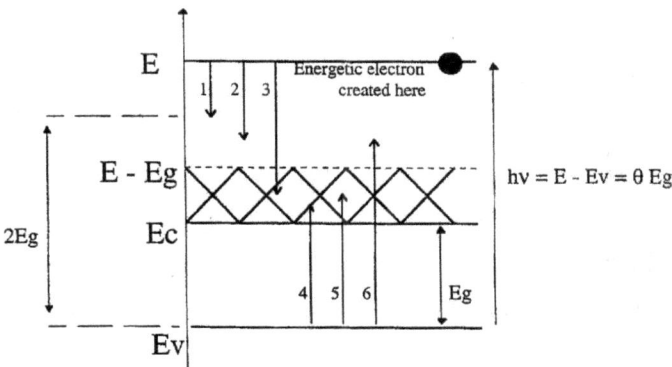

Figure 8.A.2: This illustrates a simple calculation of the probability P introduced in Section 8.2.7.

band. Hence of all possible transitions, only those ending in the cross-hatched region of the energy band diagram can lead to a (single) impact ionization. The energy delivered lies in the range $(E_g, E - E_c)$.

Similarly, of the transitions 4, 5, 6, etc., from the top of the valence band, only transitions of the type 4 and 5 absorb the relevant energy lying in the range $(E_g, E - E_g - E_v = E - E_c)$ to promote an electron into the conduction band. Again of all possible transitions, only those ending in the cross-hatched region are relevant to the single impact ionization.

The simple calculation [25] took the probability of impact ionization as the ratio of the number of favorable electron transitions (N_E) divided by the number of possible transitions M_E for a single impact ionization.

Let $g(\varepsilon) d\varepsilon$ be the number of states in the range $d\varepsilon$ in the conduction band at energy ε. Also let

$$f_0(\varepsilon) \equiv 1 - 1/[1 + \exp(\varepsilon - E_F)/kT_C]$$

be the probability of an electron vacancy at energy level ε, E_F being the Fermi level and T_C the converter temperature. Then

$$M_E = \int_{E_C}^{E} g(\varepsilon) \, d\varepsilon$$

$$N_E = \int_{E_C}^{E-E_g} g(\varepsilon) f_0(\varepsilon) \, d\varepsilon$$

It is clear from Fig. 8.A.2 that we require $E > E_g$. In fact, because of energy and momentum conservation in electron collisions, one needs to increase the minimum energy involved in impact ionization from E_g to $\theta E_g (\theta > 1)$, as discussed in Appendix 8.A.1.

Appendix 8.A.3: Note on Fig. 8.9

Because of energy and momentum conservation an electron in the conduction band with a kinetic energy E_g cannot cause an impact ionization. It needs more energy than that. The exciting photon has to have an energy θE_g The impact ionizing electron has then a kinetic energy $(\theta - 1)E_g$. In fact, it is found that in the simple case shown in Fig. 8.9 [1, p. 250]

$$\theta = 1 + (2m_c + m_h)/(m_c + m_h)$$

This has the value 2 for large $m_h \gg m_c$, which is quite possible, but it reaches 2.5 for $m_c = m_h$ and 3 for large $m_c \gg m_h$.

8.3 Discussion following Peter Landsberg's presentation

Prince: Is impact ionization a significant effect in the real world, i.e. what is its numerical probability?

Landsberg: I can't give you a precise number from memory, but it has definitely been measured in several semiconductor experiments. For example, quantum efficiencies greater than 1 are frequently observed. The probabilities are, however, relatively small. If I had to guess I'd think it would be around 20% for the generation of one additional electron.

Prince: But silicon has a band gap of 1.1 eV. Would the probability be large enough for, say, a 5 eV photon to produce two additional electrons by impact ionization?

Landsberg: In principle, I think so.

Bogus: Some years ago the Chinese claimed to have achieved 36% efficiency by fluorescing a short wave photon into two longer wave ones in order to generate 2 electrons.

Landsberg: Yes, that was Lee *et al.* The theory was sound but nobody was able to reproduce their results. I think that their paper is largely discredited these days.

Faiman: Your slide of probability, for impact ionization, versus band gap peaked at an energy of about 0.6 eV. This is very much less than the band gaps of around 1.5 eV one normally associates with high efficiency solar cells. This means that if you wanted to use impact ionization to increase the efficiency of a normal cell you would have to make a trade-off in your choice of band gap. Furthermore, whatever the probability for multi-electron generation might be for high photon energies, the flux of such photons in AM 1.5 sun light is very small (don't forget that essentially no photons with energies above 4.2 eV reach ground level because of the ozone layer), so the net effect will have to be small. For both of these reasons I find it hard to believe that this phenomenon could be employed to increase cell efficiency in any meaningful manner.

Weil: Martin Wolf, in a 1971 classic paper, tabulated all of the loss mechanisms in PV cells, according to the relative magnitudes of their effects. He found that the two largest effects were photons with energies that are substantially larger than the band gap, as you have discussed, and also long wave photons with insufficient energy for photoexcitation.

Cahen: In the late 1980s, together with Martin Wolf, I developed an experimental method for measuring virtually all the effects listed in that table. The paper was published in a special issue of IEEE circa 1989/90.

8.4   Plastic solar cells for the 21st century

Invited keynote presentation by Prof. N. Serdar Sariciftci, Christian Doppler Laboratory for Plastic Solar Cells, Physical Chemistry, Johannes Kepler University of Linz Altenbergerstr. 69, A-4040 Linz, Austria

8.4.1   *Abstract*

Recent developments on the conjugated polymer based photovoltaic elements have been reviewed. Photophysics of such photoactive devices is based on the photoinduced electron transfer from donor type semiconducting, conjugated polymers onto acceptor type conjugated polymers or acceptor molecules such as Buckminsterfullerene, C_{60}. The photoinduced electron transfer in solid composite films of fullerenes embedded into conjugated polymers is reversible, ultrafast (within 300 fs) with a quantum efficiency approaching unity, and metastable. The photophysics of conjugated polymer solar cells as well as examples of different photovoltaic architectures are discussed with their potential in terrestial solar energy conversion.

8.4.2 *Introduction*

In the last couple of years several groups reported photovoltaic elements using conjugated polymers with incident photon to collected electron (IPCE) efficiencies of around 20% and total power conversion efficiencies are around 1% for AM1.5. Despite the low efficiency, the industrial interest in this field has been propelled by the demonstration of large area (10 cm × 15 cm), flexible, low cost photovoltaic cells demonstrated in our group. It has also seen a breakthrough by the Cambridge group using the simple lamination procedure of two polymer films to achieve an industrially attractive method for producing large area solar cells.

The flexibility of chemical tailoring of desired properties, as well as the cheap technology already well developed for all kinds of plastic thin film applications make photovoltaic elements based on polymeric materials quite attractive.

An encouraging breakthrough to higher efficiencies is achieved by mixing electron-donor type polymers with suitable electron acceptors. To overcome the limitation of the photoinduced charge carrier generation, this dual molecule approach has been successful [1–4]. For example, in such devices, consisting of a composite thin film with conjugated polymer/fullerene mixture, the photogeneration efficiency of charges is near 100%. In such a single layer photoactive mixture film a "bulk heterojunction" is formed between the electron donors and acceptors (Fig. 8.10). It is the photophysics between these donor/acceptor subsystems which drives the photovoltaic activity. In the following we will briefly review these interesting phenomena in the system consisting of conjugated polymers as donors and fullerenes (and derivatives of fullerenes) as electron acceptors.

Conjugated polymers in their undoped, semiconducting state are electron donors upon photoexcitation (electrons promoted to the antibonding π^* band). Once the photoexcited electron is transferred

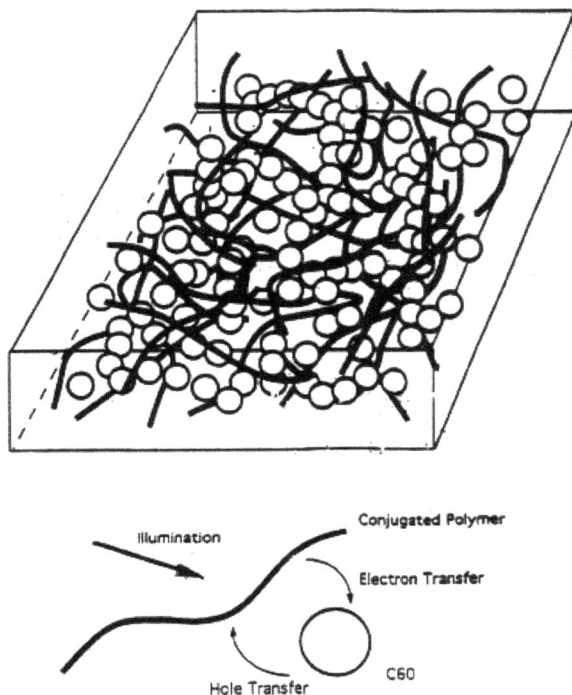

Figure 8.10:   Schematic illustration of a "bulk heterojunction".

to an acceptor unit, the resulting cation radical (positive polaron) species on the conjugated polymer backbone is known to be highly delocalized and stable as shown in electrochemical and/or chemical oxidative doping studies.

Independently, the Santa Barbara group and the Osaka group reported studies on the photophysics of mixtures of conjugated polymers with C_{60} [1, 2, 5–11]. The observations clearly evidenced an ultrafast, reversible, metastable photoinduced electron transfer from conjugated polymers onto Buckminsterfullerene in solid films. Using this molecular effect at the interface between bilayers consisting of semiconducting polymer (poly(2-methoxy,5-(2'-ethyl-hexoxy)-p-phenylene) vinylene, (hereafter referred to as MEH-PPV) and C_{60} films, diodes were demonstrated with rectification ratios on the order of 10^4 which exhibited a photovoltaic effect [2].

Significant improvement of the relatively low collection efficiency of the D/A bilayer has been achieved by using phase separated composite materials through control of the morphology of the phase separation into an interpenetrating network ("bulk hetero-junction"). Power conversion efficiency of solar cells made from MEH-PPV/C$_{60}$ composites was subsequently increased by two orders of magnitude to approximately 3% for monochromatic irradiation into the absorption band [3]. The Cambridge group with a parallel approach using acceptor type conjugated polymers in composite with MEH-PPV has also realized efficient polymeric photovoltaic devices [4].

8.4.3 *Ultrafast photoinduced electron transfer from conjugated polymers onto C$_{60}$*

The optical absorption spectrum of an MEH-PPV/C$_{60}$ film is a simple superposition of the two component's spectra without any indication of interaction between the two materials in the ground state. The strong luminescence of MEH-PPV however, is quenched by a factor in excess of 10^3 [1] indicating the existence of a rapid quenching process; for example, sub-picosecond electron transfer [12]. The strong quenching of the luminescence of another conjugated polymer poly(3-octylthiophene) (P3OT) reported by Morita *et al.* [8] is also consistent with efficient photoinduced electron transfer. Thus, the quenching of luminescence has been observed in a number of conjugated polymers in composite with fullerenes, indicating this to be a general phenomenon for the non-degenerate ground state conjugated polymers [5].

To investigate the excited state with spin selective spectroscopic methods, photoinduced absorption detected magnetic resonance experiments were performed in conjugated polymer/C$_{60}$ composites giving evidence for a complete quenching of the MEH-PPV triplet–triplet absorption signal at 1.35 eV. Instead, a strong spin $= 1/2$ signal dominates, indicating charged polarons as photoexcitations on the polymer donor [13]. This confirms that the photoinduced electron transfer occurs on a time scale sufficiently fast to quench the intersystem crossing to the triplet state.

Direct observation of the photoinduced electron transfer from conjugated polymers onto C_{60} is shown using sub-picosecond photoinduced absorption (PIA) studies. Upon adding C_{60} to P3OT, the PIA spectrum, decay kinetics, and intensity dependence all change dramatically [14–16]. At 1 ps after photoexcitation by a 100 fs pump pulse at 2.01 eV, the PIA spectrum for P3OT/C_{60} (1%) already shows a single broad, long living PIA band with virtually no evidence of the features seen in pristine P3OT at 1.9 and 1.2 eV. This ultrafast (<1 ps) formation of the PIA band at 1.55 eV again demonstrates that the electron transfer occurs on a sub-picosecond time scale.

The admixture of 1% of C_{60} into the conjugated polymer matrix results in an increase of initial photocurrent by an order of magnitude. This increase of the photocarrier generation efficiency is accompanied by the succesive increase in lifetime of the photocarriers upon adding more and more C_{60}. Thus, the ultrafast photoinduced electron transfer from the semiconducting polymer onto C_{60} not only enhances the charge carrier generation in the host polymer but also serves to prevent recombination by separating the charges and stabilizing them [17]. The composite films exhibit a remarkably enhanced photoconductivity over the broad spectral range from the near infrared to the ultraviolet. This observation is in full agreement with the photoinduced electron transfer phenomenon which leaves metastable positive polarons on the polymer backbone after the electron transfer; that is, *photodoping*.

Definitive evidence of charge transfer and long lived charge separation is obtained from light-induced electron spin resonance (LESR) experiments [1]. Upon illuminating the P3OT/C_{60} composites with light of $h\nu = E_{\pi-\pi*}$ where $E_{\pi-\pi*}$ is the energy gap of the conjugated polymer (donor), two photoinduced ESR signals can be resolved; one at $g = 2.00$ and the other at $g = 1.99$ [1]. The higher g-value line is assigned to the conjugated polymer cation (polaron) and the lower g-value line to C_{60-} anion. We have further investigated the LESR of conjugated polymer/fullerene composites on the saturation behavior of the signal. These saturation experiments show clearly that the LESR signal assigned to positive polarons on the conjugated polymer backbone (high g-value) has a different saturation behavior, for example, different relaxation mechanism compared to the fullerene anion signal which is not saturating at all (rapid relaxation on the fullerene balls) [18]). We can safely conclude from these studies that the photoinduced radicals in these polymer/fullerene composites are independent of each other and dissociated completely.

8.4.4 Conjugated polymer/C₆₀ heterojunction photodiodes

From the energy band diagram (Fig. 8.11) it is clear that the hetero-junction formed at the interface between a semiconducting polymer

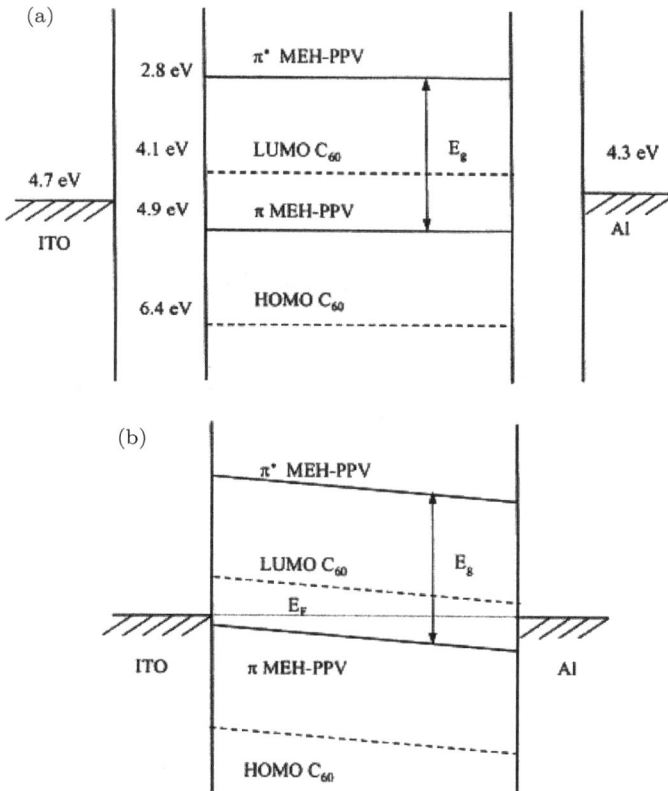

Figure 8.11:   Schematic energy band diagram of a bulk heterojunctin consisting of conjugated polymers and fullerenes; (a) open circuit, (b) short circuit condition

and a C_{60} thin film should function as a diode with a rectifying current–voltage characteristic (analogous to a p–n junction, however with a different mechanism based on molecular redox properties). In reverse bias, electron injection into the semiconducting polymer and electron removal from C_{60} are energetically unfavorable. This inherent polarity of the device results in very low reversed bias current densities. On the other hand, electron injection into C_{60} and electron removal from the semiconducting polymer are energetically favorable. Thus, resulting in relatively high current densities under forward bias. Analogous arguments for molecular diodes were first proposed by Aviram and Rattner years ago for Langmuir–Blodgett D/A structures [19]. Thus, two kinds of photodiodes have been fabricated: bilayer heterojunction devices and single layer "bulk heterojunction" diodes (Fig. 8.12).

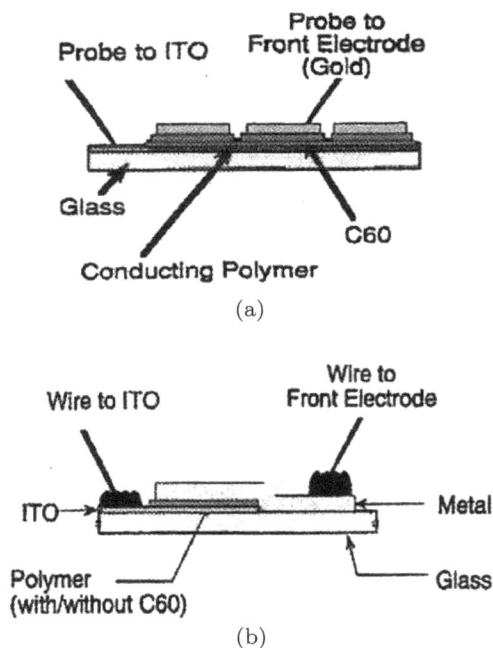

Figure 8.12: Device architectures used in the fabrication of plastic solar cells: (a) Bilayer heterojunctions, (b) single layer bult heterojunction devices.

The current–voltage characteristic of a heterojunction device consisting of successive layers of ITO/MEH-PPV/C$_{60}$/Au shows an exponential turn-on up to 500 mV in forward bias; the rectification ratio being approximately 10^4.

The current–voltage characteristic of the device changes dramatically upon illumination by visible light and exhibits photovoltaic effect with an open circuit voltage (V_{oc}) around 0.5 V and a short circuit current density (J_{sc}) of 2.08×10^{-6} A/cm^2 under 10 mW illumination, resulting in a power conversion efficiency of 0.04%.

Although the quantum efficiency for photoinduced charge separation is near unity for a D/A pair, the conversion efficiency in a bilayer heterojunction device is limited:

(i) Due to the molecular nature of the charge separation process, efficient charge separation occurs only at the D/A interface; thus, photoexcitations created far from the D/A junction recombine prior to diffusing to the heterojunction.

(ii) Even if charges are separated at the D/A interface, the photovoltaic conversion efficiency is limited by the carrier collection efficiency; that is, the separated charges must be collected with minimum losses.

8.4.5 *The "bulk heterojunction" concept*

A semiconducting polymer with asymmetric contacts (a low work function metal on one side and a high work function metal on the opposite side, metal-insulator-metal MIM diodes) functions as a "tunneling injection diode"; such devices have been described by Parker [20]. A schematic cross-sectional view of such devices is displayed in Fig. 8.12.

For photovoltaic cells made with pure conjugated polymers, energy conversion efficiencies were typically 10^{-3}–10^{-2}%, too low to be used in practical applications [21–23]. Thus, as noted above, with the addition of only 1% C$_{60}$, the photoconductivity of MEH-PPV+C$_{60}$ increases by an order of magnitude over that of pure MEH-PPV. Consequently, interpenetrating phase separated D/A network composites would appear to be ideal photovoltaic materials [3, 6].

Since any point in the composite is within a few nanometers of a D/A interface, such a composite is a bulk D/A heterojunction material (see Fig. 8.10)

Important progress has been made toward creating bulk D/A heterojunction materials [3, 4, 24, 25]. Recent reports on polymer/polymer devices with up to 4.8% power conversion efficiencies at the absorption wavelengths led to considerable improvement of the general efficiencies of these types of photovoltaic cells [25]. In their report, Granström *et al.* use a well-known technique from polymer processing which facilitates the melt-blending of two different polymer layers under applied pressure, that is, lamination [25]. The different stochiometry of the two layers creates a gradient for hole conducting and electron conducting components in the bulk heterojunction.

8.4.6   *Flexible, large area, plastic solar cells*

To fully utilize the potential of the conjugated polymers as plastic photoactive materials as well as to investigate the device technology problems of upscaling, we realized at the University of Linz, large area (6 cm × 6 cm and 15 cm × 10 cm) photovoltaic elements on flexible ITO coated plastic (PET) substrates. A picture of such an element is shown in Fig. 8.13. The schematic cross sections

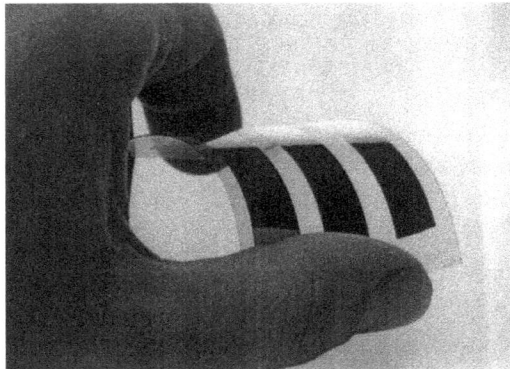

Figure 8.13:   Picture of a fully flexible, large area (6 cm × 6 cm) plastic solar cell.

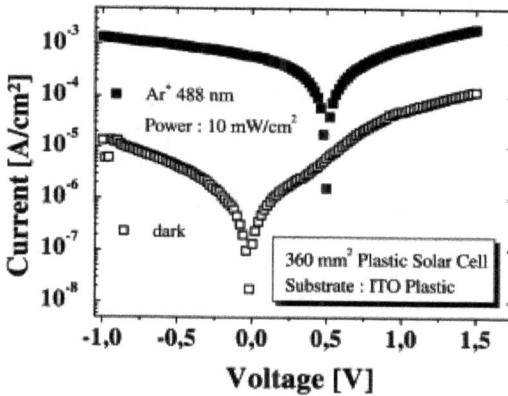

Figure 8.14: Current–voltage characteristic of the large area polymetric photovoltaic devices.

of such devices are displayed in Fig. 8.12. Devices with an open circuit potential of $\approx 0.7\,V$, a short circuit current density of $\approx 1\,mA/cm^2$ and a fill factor $FF = 0.35$ under an illumination of $10\,mW/cm^2$ at $488\,nm$ can be routinely fabricated (Fig. 8.14). The stability of such devices without any protection is extremely poor. It is imperative that the photovoltaic elements have to be protected from ambient air. The comparison of lifetimes of such a large area photovoltaic cell (fabricated in air and under arbitrary conditions without precaution towards dust-free and/or oxygen-free conditions), unprotected or protected with a special coating, clearly shows that protection against air oxygen increases the lifetime considerably.

8.4.7 *Conclusion*

The excellent photosensitivity and relatively high energy conversion efficiencies obtained from the bulk heterojunction materials are promising. Further optimization of device performance can be achieved by optimization of the device physics:

(i) Optimize the choice of metallic electrodes to achieve good ohmic contacts on both sides for collection of the oppositely charged photocarriers,

(ii) Optimize the choice of the D/A pair (the energetics influences the open circuit potential).

(iii) The band gap of the semiconducting polymer should be chosen for efficient harvesting of the solar spectrum.

(iv) Optimize the network morphology of the phase separated composite material. This way the mobility of the charge carriers within the different components of the bulk heterojunction can be maximized.

8.4.8 *Acknowledgements*

The author gratefully acknowledges his co-workers at the University of Linz, the collaboration partners J.C. Kees Hummelen and Rene A.J. Janssen (both The Netherlands), Nazario Martin (Spain), Olle Inganas (Sweden), Mats Andersson (Sweden), Michele Maggini (Italy), Maurizio Prato (Italy), David Faiman (Israel), Erhard Glotzl (Austria). Financial support from the European Commission (DGXII, JOULE III), Quantum Solar Energy Linz GesmbH (Austria), Christian Doppler Foundation, Austrian Foundation for Scientific Research (FWF P-12680CHE) as well as Netherlands Agency for Energy and Environment (NOVEM) is gratefully acknowledged.

8.4.9 *References*

[1] N. S. Sariciftci *et al.*, "Photoinduced electron transfer from conducting polymers onto buckminsterfullerene", *Science* 258 (1992) 1474.

[2] N. S. Sariciftci *et al.*, "Semiconducting polymer buckminsterfullerene heterojunctions: Diodes, photodiodes and photovoltaic cells", *Appl. Phys. Lett.* 62(6) (1993) 585.

[3] G. Yu *et al.*, "Polymer photovoltaic cells: Enhanced efficiencies via a network of internal donor-acceptor heterojunctions", *Science* 270 (1995) 1789.

[4] J. J. M. Halls *et al.*, "Efficient photodiodes from interpenetrating polymer networks", *Nature* 376 (1995), 498.

[5] N. S. Sariciftci and A. J. Heeger, "Photophysics, charge separation and device applications of conjugated polymer/fullerene composites", in H. S. Nalwa (ed.), *Handbook of Organic Conductive Molecules and Polymers*, John Wiley & Sons., 1996, pp. 414–450.

[6] N. S. Sariciftci, "Role of buckminsterfullerene in organic, photoelectric devices", *Prog. Quant. Electr.* 19 (1995) 131.

[7] N. S. Sariciftci and A. J. Heeger, U.S. Patent, 1994. No: 5,331,183, University of California, U.S.A.

[8] S. Morita, A. A. Zakhidov, and K. Yoshino, "Doping polythiophenes with fullerenes", *Sol. State Commun.* 82 (1992) 249.

[9] S. Morita *et al.* "Photoconductivity in poly3-octylthiophene with C_{60}", *Jpn. J. Appl. Phys.* 32 (1993) L1173.

[10] S. Morita *et al.*, "Doping of poly(phenylenevinylene) derivatives with C_{60}", *J. Appl. Phys.* 74(4) (1993) 2860.

[11] A. Fujii *et al.*, "Organic photovoltaic cell with donor-acceptor double heterojunctions", *Jpn. J. Appl. Phys. Pt 2* 35(11A) (1996) L1438–L1441.

[12] L. Smilowitz *et al.*, "Photoexcitation spectroscopy of conducting polymer-C_{60} composites", *Phys. Rev. B* 47 (1993) 13835.

[13] S. Wei *et al.*, "Absorption-detected magnetic-resonance studies of photoexcitations in conjugated-polymer/C-60 composites", *Phys. Rev. B-Condensed Matter* 53(5) (1996) 2187–2190.

[14] B. Kraabel *et al.*, "Ultrafast photoinduced electron transfer in conducting polymer-buckminsterfullerene composites", *Chem. Phys. Lett.* 213(3,4) (1993) 389.

[15] B. Kraabel *et al.*, "Ultrafast spectroscopy on photoinduced electron transfer from semiconducting polymer to C_{60}", *Phys. Rev. B* 50 (1994) 18543.

[16] B. Kraabel *et al.*, "Subpicosecond photoinduced electron-transfer from conjugated polymers to functionalized fullerenes", *J. Chem. Phys.* 104(11) (1996) 4267–4273.

[17] C. H. Lee *et al.*, "Sensitization of photoconductivity of conducting polymers by C_{60}", *Phys. Rev. B* 48 (1993) 15425.

[18] V. Dyakonov *et al.*, "Photoinduced charge carriers in conjugated polymer-fullerene composites studied with light-induced electron spin resonance", *Phys. Rev. B* 59 (1999) 8019.

[19] A. Aviram and M. A. Ratner, "Molecular diode", *Chem. Phys. Lett.* 29 (1974) 277.

[20] I. D. Parker, "Tunnel diodes of conjugated polymers", *J. Appl. Phys.* 75(3) (1994) 1656.

[21] G. Yu, C. Zhang, and A. J. Heeger, "Dual function of conjugated polymer devices", *Appl. Phys. Lett.* 64(12) (1994) 1540.

[22] G. Yu, K. Pakbaz, and A. J. Heeger, "High efficiency photonic devices made with semiconducting polymers", *Appl. Phys. Lett.* 64(25) (1994) 3422.

[23] H. Antoniadis *et al.*, "Photovoltaic effects at polymer/metal interface", *Polymer Preprints* 34(2) (1993) 490.

[24] G. Yu and A. J. Heeger, "Charge separation and photovoltaic conversion in polymer composites", *J. Appl. Phys.* 78 (1995) 4510.

[25] M. Granström *et al.*, "Laminated fabrication of polymeric photovoltaic diodes", *Nature* 395 (1998) 257.

8.5   Discussion following Serdar Sariciftci's presentation

Cahen: You mentioned that you had a problem making cells last more than a few hours. Considering that the CO, CN and CH bonds are intrinsically weak, how do you intend overcoming the lifetime problem? Are you going to split the spectrum in order to exclude certain wavelengths, or can you exclude oxygen, or what?

Sariciftci: The principal problem is oxygen, but it has been solved by old-fashioned engineering techniques. Philips have developed a proprietary encapsulation process that increases the working life time of polymer LEDs from 2.6 hours to 40,000 hours. By adding C_{60}, I believe that we shall be able to extend the starting lifetime even before the 10^4 amplification factor given by the encapsulation. Now 40,000 hours may not sound very impressive for a solar cell, but it represents about 9 years of daylight exposure and it is industrial companies rather than governments that are largely responsible for funding this work. They usually understand what they are investing in.

Katz: Is there any possibility of combining in a single device, a solar cell and an optical limiter, for example, for use as a "smart" window?

Sariciftci: It is indeed possible. I'd like to be able to say more about this interesting device but (to my embarrassment as a scientist) I am prevented from doing so for reasons of industrial security.

Rosenwaks: What kinds of efficiencies have been obtained for polymer solar cells and are these monochromatic or full-spectrum values?

Sariciftci: The only outdoor AM1.5 measurements that have been done were performed here at Sede Boqer, on cells that had degraded during their journey from Linz. David and his colleagues measured an efficiency of 0.6%. When the cells were fresh, however, we measured them in our laboratory using monochromatic lasers and, after the appropriate calculations, we arrived at an efficiency slightly above 1%.

Prince: I apologize that we must cut this discussion short as I have just received a note that *PETAL* (the $400\,m^2$ parabolic solar dish concentrator at Sede Boqer, Ed.) has been turned on the sun, for the first time, and we are all invited outside to see.

As a final comment, however, I am excited that you are already achieving integer efficiencies so early in the research program. I remember a joke that went the rounds during the early years of silicon cell research. People were asking how they would select the two people that would remain alive when the rest of the US is completely covered with these cells!

Afterword

This volume has presented the history of photovoltaic power plants up to the year 1999. Fortunately, however, history did not stop there. This fact is concisely stated in the words of one of the makers of that history, Vahan Garboushian, who, when asked by the editor to review his chapter, wrote: *David, I reviewed your attachment and everything is correct for that period. Of course the world moved on and did use Multijunction cells and more advanced modules and lenses. But the principal and the points discussed do still apply today. Vahan.* So, without further ado, the reader is referred to the forthcoming Volume 3 in this series, which will follow the history of photovoltaic power plants into the first part of the 21st century.